Progress in Probability

Volume 18

Seminar on Stochastic Processes, 1989

E. Çınlar
K.L. Chung
R.K. Getoor
Editors

P.J. Fitzsimmons
R.J. Williams
Managing Editors

1990

Birkhäuser
Boston · Basel · Berlin

E. Çınlar
Department of Civil Engineering and
Operations Research
Princeton University
Princeton, NJ 08544
U.S.A.

K.L. Chung
Department of Mathematics
Stanford University
Stanford, CA 94305
U.S.A.

R.K. Getoor
Department of Mathematics
University of California, San Diego
La Jolla, CA 92093
U.S.A.

P.J. Fitzsimmons
R.J. Williams
(Managing Editors)
Department of Mathematics
University of California, San Diego
La Jolla, CA 92093
U.S.A.

ISSN: 0892-063X

Printed on acid-free paper.

Softcover reprint of the hardcover 1st edition 1990

ISBN-13: 978-1-4612-8031-6 e-ISBN-13: 978-1-4612-3458-6
DOI: 10.1007/ 978-1-4612-3458-6

Prepared by the authors in camera-ready form.

9 8 7 6 5 4 3 2 1

FOREWORD

The 1989 Seminar on Stochastic Processes was held at the University of California at San Diego on March 30, 31 and April 1, 1989. This was the ninth in an annual series of meetings which provide researchers with the opportunity to discuss current work on stochastic processes in an informal and enjoyable atmosphere. Previous seminars were held at Princeton University, Northwestern University, the University of Florida and the University of Virginia. The seminar has grown over the years, with a total of seventy-five participants in 1989. Following the successful format of previous years, there were five invited lectures, delivered by K. L. Chung, D. Dawson, R. Durrett, N. Ikeda and T. Lyons, with the remainder of time being devoted to structured, but less formal, discussions on current work and problems. Several smaller groups also held workshop sessions on specific topics such as super-processes, diffusions on fractals and Harnack inequalities. The participants' interest and enthusiasm created a lively and stimulating environment for the seminar. A sample of the research discussed there is contained in this volume.

The 1989 Seminar was made possible by the support of the National Science Foundation, the National Security Agency and the University of California at San Diego. We extend our thanks to them, and to the publisher Birkhäuser Boston, for their support and encouragement. Finally, thanks go to Lynn Williams for her cheerful assistance with the seminar organization and production of this volume.

P. J. Fitzsimmons

R. J. Williams

La Jolla, 1989

LIST OF PARTICIPANTS

P. Arzberger
B. Atkinson
J. Azema
M. Bachman
M. Barlow
R. Bass
C. Bezuidenhout
R. Blumenthal
G. Brosamler
C. Burdzy
D. Burkholder
H. Cai
R. Carmona
W. Chen-Masters
K. L. Chung
E. Cinlar
M. Cranston
R. Dalang
R. Dante DeBlassie
R. Darling
D. Dawson
J. Deuschel
N. Dinculeanu
R. Durrett
E. B. Dynkin
M. Emery
S. N. Evans
N. Falkner
P. Fitzsimmons
R. K. Getoor
J. Glover
H. Heyer
K. Hoffmann
J. Horowitz
P. Hsu
N. Ikeda
O. Kallenberg
F. Knight
Y. Kwon
T. Kurtz
T. Liggett
T. Lyons
P. March
M. Marcus
P. McGill
T. Mountford
B. Oksendal
V. Papanicolaou
R. Pemantle
M. Penrose
E. Perkins
J. Pitman
L. Pitt
A. O. Pittenger
Z. Pop-Stojanovic
S. Port
P. Protter
K. M. Rao
J. Rosen
T. Salisbury
M. J. Sharpe
C. T. Shih
A. Sznitman
M. Taksar
L. Taylor
S. J. Taylor
G. Terdik
E. Toby
R. Tribe
J. Walsh
J. Watkins
S. Weinryb
R. Williams
Z. Zhao
W. Zheng

TABLE OF CONTENTS

R. F. BASS and K. BURDZY	A probabilistic proof of the boundary Harnack principle	1
J.-D. DEUSCHEL	Logarithmic Sobolev inequalities of symmetric diffusions	17
S. N. EVANS	Rescaling the vacancy of a Boolean coverage process	23
P. J. FITZSIMMONS, R. K. GETOOR and M. J. SHARPE	The Blumenthal-Getoor-McKean Theorem revisited	35
P. J. FITZSIMMONS and S. C. PORT	Local times, occupation times, and the Lebesgue measure of the range of a Levy process	59
J. HOROWITZ and R. L. KARANDIKAR	Martingale problems associated with the Boltzmann equation	75
P. HSU	Probabilistic methods in differential geometry	123
Z. MA and R. SONG	Probabilistic methods in Schrödinger equations	135
M. NAGASAWA	Stochastic variational principle of Schrödinger processes	165
B. ØKSENDAL	The high contact principle in optimal stopping and stochastic waves	177
Z. POP-STOJANOVIC and M. RAO	Continuity of solutions of Schrödinger equation	193
G. TERDIK	Stationary solutions for bilinear systems with constant coefficients	197
Z. ZHAO	Gaugeability for unbounded domains	207
P. J. FITZSIMMONS and R. K. GETOOR	Correction to: Some formulas for the energy functional of a Markov process	215

A PROBABILISTIC PROOF OF THE BOUNDARY HARNACK PRINCIPLE[1]

by

RICHARD F. BASS and KRZYSZTOF BURDZY

1. Introduction. The boundary Harnack principle may be stated as follows (cf. Jerison and Kenig (1982a), Theorem 5.25).

THEOREM 1.1. *Let D be a Lipschitz domain and V an open set. For any compact $K \subseteq V$, there exists a constant c_0 such that for all positive harmonic functions u and v in D that vanish continuously on $(\partial D) \cap V$ with $u(x) = v(x)$ for some $x \in K \cap D$,*

$$c_0^{-1} u(y) < v(y) < c_0 u(y) \quad \text{for all } y \in K \cap D.$$

The boundary Harnack principle was first proved by Dahlberg (1977). Subsequently Wu (1978) and Jerison and Kenig (1982a) gave alternate proofs. The result was extended in many directions, see, e.g., Caffarelli, Fabes, Mortola and Salsa (1981), Fabes, Garofalo and Salsa (1986), Fabes, Garofalo, Marin-Malave, and Salsa (1989) and Jerison and Kenig (1982b).

A related problem is to identify the Martin boundary for Lipschitz domains. Hunt and Wheeden (1970) showed that in a bounded Lipschitz domain the Martin boundary may be identified with the Euclidean one. Jerison and Kenig (1982a) showed how this result follows from the same techniques that they used to prove Theorem 1.1.

The main purpose of this paper is to give a probabilistic proof of Theorem 1.1, one using elementary properties of Brownian motion. We also obtain the fact

[1] Research partially supported by NSF grants DMS 8701073 and DMS 8901255.

that the Martin boundary equals the Euclidean boundary as an easy corollary of Theorem 1.1. The boundary Harnack principle may be viewed as a Harnack inequality for conditioned Brownian motion; as an application we prove some new probability bounds for conditioned Brownian motion in Lipschitz domains.

The principal motivation for this work was to give a proof of the boundary Harnack principle and of the Martin boundary result that could be easily extended to domains more general than Lipschitz: ones where locally the boundary is the graph of a continuous function with a modulus of continuity weaker than Lipschitz. See Bass and Burdzy (1989).

In Section 2 the main estimate on Brownian motion in Lipschitz domains is obtained. Theorem 1.1 is proved in Section 3, while the Martin boundary result is given in Section 4. Section 5 contains the estimates on conditioned Brownian motion.

2. The main estimate. Theorem 1.1 is essentially a local result, and for the time being we work with domains lying above the graph of a Lipschitz function. So let $\lambda > 0$ and suppose $\Gamma : \mathbf{R}^{d-1} \to \mathbf{R}$ is a bounded Lipschitz function with Lipschitz constant λ. For points $x = (x_1, \dots, x_d)$ in $\mathbf{R}^d$ we write $x = (\widetilde{x}, x_d)$, where $\widetilde{x} = (x_1 \dots, x_{d-1})$. Let

$$D = \{x \in \mathbf{R}^d : x_d > \Gamma(\widetilde{x})\}.$$

We let

$$\Delta(x,a,r) = \{y \in D : \Gamma(\widetilde{y}) < y_d < \Gamma(\widetilde{y}) + a, \ |\widetilde{y} - \widetilde{x}| < r\}, \tag{2.1}$$

$$\partial^u \Delta(x,a,r) = \{y \in \partial\Delta(x,a,r) : y_d = \Gamma(\widetilde{y} + a)\}, \qquad (\text{“}u\text{”} = \text{upper}),$$

and

$$\partial^s \Delta(x,a,r) = \{y \in \partial\Delta(x,a,r) : |\widetilde{y} - \widetilde{x}| = r\}, \qquad (\text{“}s\text{”} = \text{side}).$$

Let (X_t, P^x) be Brownian motion in $\mathbf{R}^d$. For any Borel set A, let

$$T(A) = \inf\{t : X_t \in A\}.$$

The main estimate that we obtain in this section says that the probability that Brownian motion leaves $\Delta(x,a,r)$ near the boundary of D is bounded by a constant times the probability it leaves far from ∂D. First we have

LEMMA 2.1. *There exist a constant $c_1 = c_1(\lambda) \in (0,1)$ such that*

(a) *if $a > 0$, $r \geq a$, and $y \in D$ with $\widetilde{y} = \widetilde{x}$ and $y_d \in [\Gamma(\widetilde{x}) + a/2, \Gamma(\widetilde{x}) + a]$, then*

$$P^y(T(\partial\Delta(x,a,r)) = T(\partial^u\Delta(x,a,r))) \geq c_1;$$

(b) *if $a > 0$ and $y \in \Delta(x,a,a)$ with $\widetilde{y} = \widetilde{x}$, then*

$$P^y(T(\partial\Delta(x,a,a)) = T(\partial^s\Delta(x,a,a))) \leq 1 - c_1; \qquad \text{and}$$

(c) *if $k \in \mathbb{Z}^+$, $a > 0$, $r \geq ak$, and $y \in \Delta(x,a,r)$ with $\widetilde{y} = \widetilde{x}$, then*

$$P^y(T(\partial\Delta(x,a,r)) = T(\partial^s\Delta(x,a,r))) \leq (1 - c_1)^k.$$

Proof: The proof is elementary. By scaling we may suppose $a = 1$. Choose $c_2 = (\lambda^{-1} \wedge 1)/8$. Let

$$J_1 = \{y : |\widetilde{y} - \widetilde{x}| < c_2,\ \Gamma(\widetilde{x}) + \frac{1}{4} < y_d < \Gamma(\widetilde{x}) + 2\},$$
$$J_2 = \{y : |\widetilde{y} - \widetilde{x}| < c_2,\ \Gamma(\widetilde{x}) - 2 < y_d < \Gamma(\widetilde{x}) + 2\},$$

and

$$\partial^u J = \{y : |\widetilde{y} - \widetilde{x}| < c_2,\ y_d = \Gamma(\widetilde{x}) + 2\}.$$

It is easy to see that there exists c_1 depending only on c_2 such that

$$P^y(T(\partial J_1) = T(\partial^u J)) \geq c_1 \qquad \text{if} \quad \widetilde{y} = \widetilde{x},\ y_d \in (\Gamma(\widetilde{x}) + \frac{1}{2}, \Gamma(\widetilde{x}) + 1)$$

and

$$P^y(T(\partial J_2) = T(\partial^u J)) \geq c_1 \qquad \text{if} \quad \widetilde{y} = \widetilde{x},\ y_d \in (\Gamma(\widetilde{x}), \Gamma(\widetilde{x}) + 1).$$

Note that if $T(\partial J_1) = T(\partial^u J)$, then $T(\partial\Delta(x,1,r)) = T(\partial^u\Delta(x,1,r))$ P^y-a.s. for y such that $\widetilde{y} = \widetilde{x}$, $y_d \in (\Gamma(\widetilde{x}) + \frac{1}{2}, \Gamma(\widetilde{x}) + 1)$; this proves (a). Similarly, if $T(\partial J_2) = T(\partial^u J)$, then $T(\partial\Delta(x,1,1)) \neq T(\partial^s\Delta(x,1,1))$, which proves (b).

Part (c) follows from part (b) by use of the strong Markov property. Using (b),

$$\begin{aligned}P^y(T(\partial\Delta(x,1,k)) = T(\partial^s\Delta(x,1,k)))\\ \leq E^y(P^{X(U)}(T(\partial\Delta(X(U),1,1)) = T(\partial^s\Delta(X(U),1,1));\\ T(\partial\Delta(x,1,k-1)) = T(\partial^s\Delta(x,1,k-1)))\\ \leq (1-c_1)P^y(T(\partial\Delta(x,1,k-1)) = T(\partial^s\Delta(x,1,k-1))),\end{aligned}$$

where

$$U = T(\partial\Delta(x,1,k-1)).$$

Using induction completes the proof. □

Let

$$F_1 = \{T(\partial\Delta(0,3,3)) = T(\partial^s\Delta(0,1,3))\}.$$

Let

$$\partial^g\Delta(0,3,3) = \partial\Delta(0,3,3) \setminus (\partial D \cup \partial^s\Delta(0,1,3)), \qquad (\text{“}g\text{”} = \text{good}).$$

Let

$$F_2 = \{T(\partial\Delta(0,3,3)) = T(\partial^g\Delta(0,3,3))\}.$$

The main result of this section is

THEOREM 2.2. *There exists $c_3 = c_3(\lambda) < \infty$ such that for all $x \in \Delta(0,3,1)$,*

$$P^x(F_1) \leq c_3 P^x(F_2).$$

Proof: Choose $M \in \mathbf{Z}^+$ so that $(1-c_1)^M < c_1/2$ and $M \geq 2\sum_{i=1}^{\infty} i2^{-i}$, where c_1 is the constant of Lemma 2.1. Let

$$J_k = \{y \in D : y_d \in [\Gamma(\tilde{y}) + M^{-2}2^{-k-1}, \Gamma(\tilde{y}) + M^{-2}2^{-k}],\ |\tilde{y}| \leq 2 - M^{-1}\sum_{i=1}^{k} i2^{-i}\}.$$

Arguing just as in the proof of Lemma 2.1(a), there exists a constant $c_4 = c_4(\lambda) \in (0,1)$ such that

(2.2) $$P^z(F_2) \geq c_4, \qquad z \in \Delta(0,3,2) \setminus \Delta(0, M^{-2}/4, 2).$$

Our first goal is to prove that

(2.3) $$P^z(F_2) \geq c_4 c_1^{m-1} \quad \text{for } z \in J_m.$$

We use induction. By (2.2), we have that (2.3) holds for $m = 1$. Suppose (2.3) holds for m, and suppose $z \in J_{m+1}$. For the remainder of the proof, write

(2.4) $$\Delta_m = \Delta(z, M^{-2}2^{-m}, mM^{-1}2^{-m}), \quad \text{and} \quad U_m = T(\partial\Delta_m).$$

By the strong Markov property,

$$P^z(F_2) \geq E^z(P^{X(U_m)}(F_2); X(U_m) \in \partial^u\Delta_m).$$

Since $\partial^u\Delta_m \in J_m$ when $z \in J_{m+1}$ and since $m \geq 1 \geq M^{-1}$, then by Lemma 2.1 (a) and the induction hypothesis,

$$P^z(F_2) \geq c_4 c_1^{m-1} P^z(X(U_m) \in \partial^u\Delta_m) \geq c_4 c_1^m.$$

So (2.3) is proved.

Now let

$$d_m = \sup_{z \in J_m} P^z(F_1)/P^z(F_2).$$

By (2.2),

$$P^z(F_1) \leq 1 \leq c_4^{-1} P^z(F_2), \qquad z \in \Delta(0,3,2) \setminus \Delta(0, M^{-2}/4, 2).$$

Hence $d_1 < \infty$, and so to prove the theorem, it suffices to prove that $\sup_m d_m < \infty$, since $\Delta(0, M^{-2}/2, 1) \subset \bigcup_{m=1}^\infty J_m$.

Consider $z \in J_{m+1}$. Using the strong Markov property, we have

$$(2.5) \qquad P^z(F_1) \le E^z(P^{X(U_m)}(F_1);\ X(U_m) \in \partial^u \Delta_m) + P^z(X(U_m) \in \partial^s \Delta_m)$$

and

$$(2.6) \qquad P^z(F_2) \ge E^z(P^{X(U_m)}(F_2);\ X(U_m) \in \partial^u \Delta_m).$$

Since $\partial^u \Delta_m \subseteq J_m$, the definition of d_m says that the first term on the right of (2.5) is bounded by

$$d_m E^z(P^{X(U_m)}(F_2);\ X(U_m) \in \partial^u \Delta_m) \le d_m P^z(F_2).$$

By Lemma 2.1 (c), the second term on the right of (2.5) is bounded by

$$(1-c_1)^{mM} \le (c_1/2)^m \le 2^{-m} c_4^{-1} P^z(F_2),$$

using (2.3).

Hence, substituting in (2.5),

$$P^z(F_1) \le (d_m + 2^{-m} c_4^{-1}) P^z(F_2).$$

Thus $d_{m+1} \le d_m + c_4^{-1} 2^{-m}$, or $\sup_m d_m \le d_1 + c_4^{-1} \sum_{m=1}^{\infty} 2^{-m} < \infty$ as required.
□

3. Boundary Harnack principle. We first borrow an elementary lemma from Jerison and Kenig (1982a), Lemma 5.4. The notation is as in Section 2.

LEMMA 3.1. *There exists a constant $c_5 = c_5(\lambda) > 0$ such that if u is positive and harmonic in $\Delta(x,5,5)$ and vanishes continuously on $\Delta(x,5,5) \cap \partial D$, where $x_d = \Gamma(\widetilde{x}) + 1$, then u is bounded in $\Delta(x,3,3)$ by $c_5 u(x)$.*

Proof: Fix x. Without loss of generality, assume $u(x) = 1$. Let

$$J_k = \Delta(x, 2^{-k}, 4) \setminus \Delta(x, 2^{-k-1}, 4), \qquad k = 1, 2, \ldots .$$

By the usual Harnack inequality, u is bounded in $\Delta(x,4,4)\backslash\bigcup_{k=2}^{\infty} J_k$ by a constant $c_6 = c_6(\lambda)$.

If u is harmonic and positive in $\Delta(y,1,1)$ and $y_d = \Gamma(\widetilde{y}) + \frac{1}{2}$, then by the usual Harnack inequality there exists $c_7 = c_7(\lambda) > 0$ so that u is bounded on $\Delta(y,1,1) \setminus \Delta(y,\frac{1}{4},1)$ by $c_7 u(y)$.

Using this observation together with scaling, we see that

$$\sup_{J_{k+1}} u \leq c_7 \sup_{J_k} u,$$

and hence

(3.1) $$\sup_{J_k} u \leq c_6 c_7^k.$$

This implies that there exist constants $c_8 = c_8(\lambda), \beta = \beta(\lambda) > 0$ such that if

$$r(y) = y_d - \Gamma(\widetilde{y}),$$

then

(3.2) $$r(y) \leq c_8 u(y)^{-\beta}.$$

Suppose $y \in \Delta(x,3,3)$. Arguing as in Lemma 2.1 (a), there is a constant $c_9 = c_9(\lambda) > 0$ such that

$$P^y(T(\partial\Delta(y,2r(y),2r(y)) = T(\partial D)) \geq c_9.$$

Now let $M = (1-c_9)^{-1}$ and let N be a large real to be chosen later. Suppose there exists $x^{(1)} \in \Delta(x,3,3)$ with $u(x^{(1)}) \geq NM$. We now show that this implies there exist $x^{(2)}, \ldots, x^{(n)}, \ldots \in \Delta(x,4,4)$ with $u(x^{(n)}) \geq NM^n$, $x^{(k+1)} \in \Delta(x^{(k)}, 3r_k, 3r_k)$, where $r_k = r(x^{(k)})$. We use induction. Suppose we have $x^{(1)}, x^{(2)}, \ldots, x^{(n)}$.

Write Δ_n for $\Delta(x^{(n)}, 2r_n, 2r_n)$. Note

$$\begin{aligned} u(x^{(n)}) = E^{x^{(n)}} u(X_{T(\partial\Delta_n)}) &\leq (\sup_{\partial\Delta_n} u) P^{x^{(n)}}(T(\partial\Delta_n) \neq T(\partial D)) \\ &\leq (\sup_{\partial\Delta_n} u)(1 - c_9). \end{aligned}$$

Hence there exists $x^{(n+1)} \in \partial\Delta_n \subseteq \Delta(x^{(n)}, 3r_n, 3r_n)$ with

$$u(x^{(n+1)}) \geq (1 - c_9)^{-1} u(x^{(n)}) \geq NM^{n+1}.$$

By (3.2),

$$r_{n+1} \leq c_8(NM^{n+1})^{-\beta},$$

and so provided we take N sufficiently large so that $\sum_{i=1}^{\infty} c_8(NM^i)^{-\beta} < \frac{1}{4}$, then $x^{(n+1)} \in \Delta(x, 4, 4)$.

We thus have our sequence $x^{(n)}$ in $\Delta(x, 4, 4)$ with $u(x^{(n)}) \to \infty$. Moreover, by (3.2), $r_n \to 0$. But this contradicts the assumption that u vanishes continuously on $(\partial D) \cap \Delta(x, 5, 5)$. So we must have u bounded on $\Delta(x, 3, 3)$ by NM. □

We now prove the following special case of the boundary Harnack principle.

THEOREM 3.2. *There exists a constant $c_{10} = c_{10}(\lambda) > 0$ such that if $x \in D$ with $x_d = \Gamma(\tilde{x}) + 1$, u and v are positive and harmonic on $\Delta(x, 5, 5)$, vanish continuously on $\partial D \cap \Delta(x, 5, 5)$, and $u(x) = v(x) = 1$, then*

$$c_{10}^{-1} u(y) < v(y) < c_{10} u(y) \quad \text{for all } y \in \Delta(x, 3, 1).$$

Proof: Recall the definitions of F_1 and F_2 of Theorem 2.2. By Lemma 3.1, u is bounded on $\Delta(x, 3, 3)$ by c_5. Then if $y \in \Delta(x, 3, 1)$,

$$\begin{aligned} (3.3) \qquad u(y) = E^y u(X_{T(\partial\Delta(x,3,3))}) &\leq c_5 P^y(T(\partial\Delta(x, 3, 3)) \neq T(\partial D)) \\ &\leq c_5(P^y(F_1) + P^y(F_2)) \\ &\leq c_5(1 + c_3) P^y(F_2) \end{aligned}$$

by Theorem 2.2.

On the other hand, by the usual Harnack inequality, there exists $c_{11} = c_{11}(\lambda) > 0$ such that v is bounded below by c_{11} on $\partial\Delta(0,3,3)\backslash(\partial D \cup \partial^s\Delta(0,1,3))$. Then

$$v(y) = E^y v(X_{T(\partial\Delta(x,3,3))}) \geq c_{11} P^y(F_2). \tag{3.4}$$

Comparing (3.3) and (3.4) gives the left hand inequality, and reversing the roles of u and v gives the right hand inequality. □

However, Theorem 3.2 is actually equivalent to Theorem 1.1. We first recall the definition of a Lipschitz domain.

A bounded domain D is a Lipschitz domain if for each $x \in \partial D$ there is a Lipschitz function $\Gamma_x : \mathbf{R}^{d-1} \to \mathbf{R}$, a coordinate system CS_x, and $r_x > 0$ such that if $y = (y_1, \ldots, y_d)$ in CS_x coordinates, then

$$D \cap B(x, r_x) = \{y : y_d > \Gamma_x(y_1, \ldots, y_{d-1})\} \cap B(x, r_x).$$

Proof of Theorem 1.1: Theorem 1.1 follows from using scaling, Theorem 3.2 and the usual Harnack principle repetitively. □

4. Martin boundary. In this section we prove that the Martin boundary of a Lipschitz domain may be taken to be the Euclidean boundary. For details about Martin boundary, see Doob (1984).

Suppose D is a bounded Lipschitz domain. We denote the Green function for D by $G(x,y)$.

Let us fix $x_0 \in D$ and suppose $\varepsilon < \operatorname{dist}(x_0, \partial D)/4$.

LEMMA 4.1. *Suppose $x \in D$ with $|x - x_0| > 4\varepsilon$. There exists a constant $c_{12}(\varepsilon, D, x_0, x)$ such that*

$$G(x,y)/G(x_0,y) \leq c_{12} \quad \text{for } y \text{ in } D \setminus (\overline{B(x_0,\varepsilon)} \cup \overline{B(x,\varepsilon)}).$$

Moreover $c_{12}(\varepsilon, D, x_0, x) \to 0$ uniformly as $\operatorname{dist}(x, \partial D) \to 0$.

Proof: Pick $y_0 \in \partial B(x_0, 2\varepsilon)$. If G^0 is the Green function for Brownian motion killed on exiting $\partial B(x_0, 3\varepsilon)$, then

$$G(x_0, y_0) \geq G^0(x_0, y_0) \geq \delta(\varepsilon) > 0.$$

(See Section 1.11 of Durrett (1984) for an explicit expression for G^0.)

On the other hand, $G(x, y_0)$ is bounded above by the Newtonian potential evaluated at x, y_0; hence $G(x, y_0)$ is bounded above by a constant depending on ε (use the Green function for a large ball containing D instead of the Newtonian potential in the case $d = 2$). Moreover $G(x, y_0) \to 0$ uniformly as $\text{dist}(x, \partial D) \to 0$.

Thus the ratio $G(x, y_0)/G(x_0, y_0)$ is bounded above. But Theorem 1.1 says that $G(x, y)/G(x_0, y)$ is comparable to $G(x, y_0)/G(x_0, y_0)$ for all points y in $D \setminus (B(x, \varepsilon/2) \cup B(x_0, \varepsilon/2))$; here $u = G(x, \cdot), v = G(x_0, \cdot)G(x, y_0)/G(x_0, y_0)$. The lemma follows. □

We now prove that for fixed x_0, x, the ratio $G(x, y)/G(x_0, y)$ is Hölder continuous in y.

LEMMA 4.2. *Let x, x_0, ε be as above. Then $G(x, y)/G(x_0, y)$ is a Hölder continuous function of y for $y \in D \setminus (\overline{B(x_0, \varepsilon)} \cup \overline{B(x, \varepsilon)})$, the constants depending only on x, x_0, ε, and D.*

Proof: For a set A, define

$$\operatorname*{Osc}_A f = \sup_A f - \inf_A f.$$

Let $f(y) = G(x, y)/G(x_0, y)$. Let $y_0 \in D_\varepsilon = D \setminus (\overline{B(x_0, \varepsilon)} \cup \overline{B(x, \varepsilon)})$. Since f is bounded by c_{12} on the region D_ε by Lemma 4.1, then $\operatorname*{Osc}_{D_\varepsilon} f \le c_{12}$. So to prove the lemma, it will suffice to show that there exists $\rho = \rho(D, \varepsilon, x, x_0) < 1$ such that

$$\operatorname*{Osc}_{D \cap B(y_0, r)} f \le \rho \operatorname*{Osc}_{D \cap B(y_0, 2r)} f, \qquad r < \varepsilon/4. \tag{4.1}$$

Suppose $r < \varepsilon/4$, and let g be the ratio of any two positive harmonic functions on $D_{\varepsilon/4}$ vanishing continuously on ∂D. By considering $ag + b$ for suitable a and b, we may assume

$$\sup_{D \cap B(y_0, 2r)} g = 1, \qquad \inf_{D \cap B(y_0, 2r)} g = 0.$$

If $\sup_{D\cap B(y_0,r)} g \le \frac{1}{2}$, then since $\inf_{D\cap B(y_0,r)} g \ge 0$,

$$\operatorname*{Osc}_{D\cap B(y_0,r)} g \le \frac{1}{2}.$$

If $\sup_{D\cap B(y_0,r)} g \ge \frac{1}{2}$, there exists a point y_1 in $D \cap B(y_0, r)$ with $g(y_1) \ge \frac{1}{2}$. But then by Theorem 1.1, there exists a constant $c_{13} = c_{13}(\varepsilon, D, x, x_0) \in (0, 1)$ such that

$$\inf_{D\cap B(y_0,r)} g \ge c_{13} g(y_1).$$

Since $\sup_{D\cap B(y_0,r)} g \le 1$, in this case we have

$$\operatorname*{Osc}_{D\cap B(y_0,r)} g \le 1 - c_{13}/2.$$

Since $\operatorname*{Osc}_{D\cap B(y_0,2r)} g = 1$, we have (4.1) with $\rho = \max(\frac{1}{2}, 1 - c_{13}/2)$. □

To construct the Martin boundary of a domain, one first compactifies D by adding all limit points of the ratios $G(x,y)/G(x_0,y)$ as $y \to z$, $z \in \partial D$ (see Doob(1984)). But Lemmas 4.1 and 4.2 show that $G(x,y)/G(x_0,y)$ converges to a single value as $y \to z \in \partial D$. Thus we have proved

THEOREM 4.3. *The Martin boundary of a Lipschitz domain may be identified with a subset of the Euclidean boundary.*

To complete the identification of the Martin boundary, one needs to show that a proper subset of the Euclidean boundary will not suffice. We will write

$$K(x,z) = \lim_{y\in D, y\to z} G(x,y)/G(x_0,y) \qquad \text{for } x \in D,\ z \in \partial D.$$

We will also show that $K(x,z)$ is a minimal harmonic function for each $z \in \partial D$; that is, if u is harmonic in D satisfying $u(x) \le K(x,z)$ for all $x \in D$, then $u = cK(\,\cdot\,, z)$ for some constant c.

Theorem 4.4. *The Martin boundary of a Lipschitz domain may be identified with the Euclidean boundary.*

Proof: We first show that if $w \in \partial D$, then $K(x,w) \to 0$ uniformly as $\operatorname{dist}(x, \partial D \setminus B(x, 2\varepsilon)) \to 0$. To see this, pick $y_0 \in D \cap B(w,\varepsilon)$. Let $\delta > 0$. By Lemma 4.1, we can make $G(x,y_0)/G(x_0,y_0) < \delta$ by taking $\operatorname{dist}(x, \partial D \setminus B(w, 2\varepsilon))$ small enough. By Theorem 1.1,

$$G(x,y)/G(x_0,y) \le c_0\delta, \qquad y \in D \cap B(w,\varepsilon).$$

Now let $y \to w$ to get $K(x,w) \le c_0\delta$.

Suppose that $K(\cdot\,,w) = K(\cdot\,,z)$ for some $w,z \in \partial D$, $w \neq z$, and let $\varepsilon = |w-z|/8$. By the above argument, we have $K(x,w) \to 0$ uniformly when $\operatorname{dist}(x, \partial D \setminus B(w,2\varepsilon)) \to 0$ and when $\operatorname{dist}(x, \partial D \setminus B(z,2\varepsilon)) \to 0$. Thus, $K(x,w) \to 0$ uniformly as $\operatorname{dist}(x,\partial D) \to 0$. By the maximum principle, the positive harmonic function $K(\cdot\,,w)$ vanishes on D, contrary to the fact that $K(x_0,w) = 1$. This contradiction shows that the Martin kernels corresponding to w and z are distinct. □

Theorem 4.5. *If $z \in \partial D$, $K(\cdot\,,z)$ is a minimal harmonic function.*

Proof: Fix $z \in \partial D$ and suppose $u \le K(\cdot\,,z)$, where u is positive and harmonic. By Theorem 4.3, it follows that

$$u(\cdot) = \int K(\cdot\,,w)\mu(dw)$$

for some measure μ on ∂D. If μ is not a multiple of point mass at z, then there exists a finite measure $\widehat{\mu} \le \mu$ such that $\operatorname{dist}(z, \operatorname{supp}(\widehat{\mu})) > 0$. Let

$$\widehat{u}(\cdot) = \int K(\cdot\,,w)\widehat{\mu}(dw).$$

Then $\widehat{u}$ is positive, harmonic, and bounded by $K(\cdot\,,z)$.

Recall from the proof of Theorem 4.4 that $K(x,z) \to 0$ uniformly as $\operatorname{dist}(x, \partial D \setminus B(z,\varepsilon)) \to 0$. So the same is true of $\widehat{u}$. But for each $w \in \operatorname{supp}(\widehat{u})$, we see that $K(x,w) \to 0$ uniformly as $\operatorname{dist}(x, \partial D \cap B(z,2\varepsilon)) \to 0$ provided $2\varepsilon < \operatorname{dist}(z, \operatorname{supp}(\widehat{\mu}))$. So it follows by dominated convergence that $\widehat{u}(x) \to 0$ as $\operatorname{dist}(x, \partial D \cap B(z,2\varepsilon)) \to 0$. But then $\widehat{u}$ is a positive harmonic function vanishing continuously on ∂D, or $\widehat{u}$ is identically 0. This implies that $\widehat{\mu}$ is 0, or that μ must be point mass at z. □

5. Conditioned Brownian motion. Let h be a positive harmonic function on D and let (X_t, P_h^x) be conditioned Brownian motion, that is, the h path transform of Brownian motion. See Doob (1984) for more information about conditioned Brownian motion. In this section we prove the analog of Theorem 2.2 for (X_t, P_h^x) and we obtain an exponential bound on $P_h^x(\sup_{s\le t} |X_s - x| \ge r)$ similar to the bound for Brownian motion. We suppose we are in the set-up of Section 2 where D is the region above the graph of a Lipschitz function. The definition of F_1 and F_2 are as in Theorem 2.2.

THEOREM 5.1. *Suppose h is positive, harmonic in $\Delta(0,5,5)$, and h vanishes continuously on $\partial D \cap \Delta(0,5,5)$. Then there exists $c_{14} = c_{14}(\lambda) > 0$ such that for all $x \in \Delta(0,3,1)$*

$$P_h^x(F_1) \le c_{14} P_h^x(F_2).$$

Proof: Let z be such that $\widetilde{z} = 0$ and $z_d = \Gamma(\widetilde{z}) + 1$. Since multiplying h by a constant does not change P_h^x, let us assume $h(z) = 1$. Let $u(x) = P_h^x(F_1), v(x) = P_h^x(F_2)$. By the usual Harnack inequality, there exists $c_{15} = c_{15}(\lambda)$ such that $h \ge c_{15}$ on $\partial^g \Delta(0,3,3) = \partial \Delta(0,3,3) \setminus (\partial D \cup \partial^s \Delta(0,1,3))$. As in Lemma 2.1 (a), there exists $c_{16} = c_{16}(\lambda)$ such that $P^z(F_2) \ge c_{16}$. Then using basic properties of h path transforms,

$$v(z) = E^z(h(X_{T(\partial^g \Delta(0,3,3))}); F_2)/h(z) \ge c_{15}c_{16}.$$

Since $u(z) \le 1$, then $u(z) \le (c_{15}c_{16})^{-1} v(z)$.

The functions uh and vh are positive and harmonic (with respect to P^x) on $\Delta(0,3,3)$ and vanish continuously on $\partial D \cap \Delta(0,3,3)$ since u and v are bounded being probabilities. By the boundary Harnack principle, there exists c_{14} so that

$$(uh)(x) \le c_{14}(vh)(x) \quad \text{for } x \in \Delta(0,3,1).$$

Dividing both sides by $h(x)$ proves the theorem. □

We now obtain the following exponential bound

THEOREM 5.2. *Suppose h is as in Theorem 5.1. Then there exist $r_0 = r_0(\lambda) > 0$, $c_{17} = c_{17}(\lambda) > 0$ and $c_{18} = c_{18}(\lambda) > 0$ such that*

$$P_h^x(\sup_{s\le t} |X_s - x| > r) \le c_{17} \exp(-r^2/c_{18}t), \qquad r < r_0.$$

Proof: Since $P_h^x(F_1) + P_h^x(F_2) = 1$, then by Theorem 5.1

$$P_h^x(F_2) \ge (1 + c_{14})^{-1} \qquad \text{for } x \in \Delta(0,3,1). \tag{5.1}$$

Define

$$\tau_r = \inf\{t : |X_t - X_0| \ge r\}.$$

We have

$$P^y(\tau_\beta > 1) \ge c_{19}$$

for a constant $c_{19} = c_{19}(\lambda, \beta) > 0$. We have assumed in the proof of Theorem 5.1 that $h(z) = 1$. It follows that h is bounded above and below by constants depending only on λ on the set $\Delta(0,4,4) \setminus \Delta(0,\frac{1}{2},4)$ and

$$P_h^y(\tau_\beta > 1) \le E^y(h(X_1);\ \tau_\beta > 1)/h(y) \ge c_{20}, \qquad y \in \partial^g \Delta(0,3,3), \tag{5.2}$$

for a constant $c_{20} = c_{20}(\lambda, \beta) > 0$ provided β is taken small enough so that

$$\text{dist}\,(\partial\Delta(0,1/2,4), \partial^g\Delta(0,3,3)) \ge 2\beta.$$

So by the strong Markov property, (5.1), and (5.2),

$$P_h^x(T(\partial\Delta(0,4,4)) > 1) \ge (1 + c_{14})^{-1} c_{20}, \qquad x \in \Delta(0,3,1).$$

By scaling and the fact that $y_d - \Gamma(\tilde{y})$ is comparable to dist $(y, \partial D)$ for $y \in D$, we then get the existence of constants $c_{21} = c_{21}(\lambda) > 0$ and $p = p(\lambda) \in (0,1)$ such that

$$P_h^x(\tau_1 \le c_{21}) \le p. \tag{5.3}$$

Without loss of generality we may assume $c_{21} \leq 1$.

Let n be a positive integer to be chosen later. Let $U_1 = \tau_{1/n}, U_{i+1} = U_i + \tau_{1/n} \circ \theta_{U_i}$, where θ is the usual shift operator. Clearly $U_n \leq \tau_1$.

By (5.3) and scaling, $P_h^x(\tau_{1/n} \leq c_{21} n^{-2}) \leq p$. Hence

$$P_h^x(U_{n+1} - U_n \leq z \mid U_1, \ldots, U_n) \leq \begin{cases} p & \text{if} \quad z \leq c_{21} n^{-2} \\ 1 & \text{if} \quad z > c_{21} n^{-2} \end{cases}$$
$$\leq p + (1-p) z n^2 / c_{21} \quad \text{if} \quad z > 0.$$

By Barlow and Bass (1989), Lemma 1.1, then

$$P_h^x(\tau_1 < z) = \exp(a n^{3/2} z^{1/2} - bn),$$

where

$$a = 2((1-p)/p c_{21})^{1/2} \quad \text{and} \quad b = \log(\frac{1}{p}).$$

Taking n to be the integer part of $4b^2/9a^2 z$, for z sufficiently small we get

$$P_h^x(\tau_1 < z) \leq \exp(-c_{22}/z), \tag{5.4}$$

$c_{22} = c_{22}(\lambda) > 0$.

Using (5.4) and scaling gives Theorem 5.2, provided we take c_{17} sufficiently large. □

REFERENCES

[1] M.T. Barlow and R.F. Bass, *The construction of Brownian motion on the Sierpiński carpet*, Ann. de l'I. H. Poincaré (1989). (to appear)

[2] R.F. Bass and K. Burdzy, *The boundary Harnack principle for Hölder domains.* (in preparation)

[3] L. Caffarelli, E. Fabes, S. Mortola and S. Salsa, *Boundary behavior of non-negative solutions of elliptic operarors in divergence form*, Indiana Univ. Math. J. **30** (1981), 621–640.

[4] B. DAHLBERG, *Estimates of harmonic measure*, Arch. Rat. Mech. Anal. **65** (1977), 275–288.

[5] J.L. DOOB, "Classical Potential Theory and Its Probabilistic Counterpart," Springer, New York, 1984.

[6] R. DURRETT, "Brownian Motion and Martingales in Analysis," Wadsworth, Belmont CA, 1984.

[7] E. FABES, N. GAROFALO, S. MARIN-MALAVE, AND S. SALSA, *Fatou theorems for some nonlinear elliptic equations.* (preprint)

[8] E. FABES, N. GAROFALO AND S. SALSA, *A backward Harnack inequality and Fatou theorem for non-negative solutions of parabolic equations*, Illinois J. Math. **30** (1986), 536–565.

[9] R.R. HUNT AND R.L. WHEEDEN, *Positive harmonic functions on Lipschitz domains*, Trans. Amer. Math. Soc. **147** (1970), 507–527.

[10] D.S. JERISON AND C.E. KENIG, *Boundary value problems on Lipschitz domains*, in: Studies in Partial Differential Equations, ed. W. Littman, Washington, D.C.: Math. Assoc. Amer. (1982a).

[11] D.S. JERISON AND C.E. KENIG, *Boundary behavior of harmonic functions in non-tangentially accessible domains*, Adv. in Math. **46** (1982b), 80–147.

[12] J.-M.G. WU, *Comparisons of kernel functions, boundary Harnack principle and relative Fatou theorem on Lipschitz domains*, Ann. Inst. Fourier Grenoble **28** (1978), 147–167.

Richard F. Bass and Krzysztof Burdzy
Department of Mathematics
University of Washington
Seattle, WA 98195

Logarithmic Sobolev Inequalities of Symmetric Diffusions

BY

Jean-Dominique Deuschel

Let E be a Polish space, $\mathbf{M}_1(E)$ be the space of probability measures on E and $\mathcal{B}(E;\mathbf{R})$ be the space of bounded measurable real valued functions on E. For a given $\mu \in \mathbf{M}_1(E)$, we write $< f >_\mu = \int_E f\,d\mu$ and $\|f\|_p = \|f\|_{L^p(\mu)}$. Let $\{P_t : t > 0\}$ be a μ-symmetric Markov semigroup on E with generator L. We will suppose that the domain of L contains an algebra $\mathcal{A} \subseteq \mathcal{B}(E;\mathbf{R})$ dense in $L^p(\mu)$, $1 \le p < \infty$, which is closed under L, P_t and the composition of C^∞-functions. Moreover the semigroup should be sufficiently mixing so that for all $f \in \mathcal{A}$,

$$\lim_{t\to\infty} P_t f = < f >_\mu \qquad \text{in } L^p(\mu),\ 1 \le p < \infty.$$

We will assume that $\{P_t : t > 0\}$ is a diffusion semigroup ie. that the corresponding process can be constructed on $\Omega \equiv C\big([0,\infty);E\big)$, and we will denote by $X_t : \Omega \longmapsto E$ the evaluation map and by $\{P_x : x \in E\} \in \mathbf{M}_1(\Omega)$ the associated Markov family.

In various questions about the ergodic properties of the diffusion determined by L, an important role is played by the **logarithmic Sobolev inequality**

$$\text{(L-S)} \qquad \langle f^2 \log(f^2)\rangle_\mu - \|f\|_2^2 \log(\|f\|_2^2) \le -4\alpha \langle f \cdot Lf\rangle_\mu, \qquad f \in \mathcal{A},$$

where $\alpha \in (0,\infty)$. By L. Gross' well known result, (L-S) is equivalent to the following **hypercontractive** property of the semigroup:

$$\|P_t f\|_{q(t,p)} \le \|f\|_p, \qquad t \in (0,\infty),\ \text{for } f \in \mathcal{A} \text{ and } p \in (1,\infty),$$

where $q(t,p) = 1 + (p-1)e^{t/\alpha}$, cf. [**5**]. Moreover since we are dealing with a diffusion, this is also equivalent to the following exponential decay of the relative entropy of the process:

$$\mathbf{H}(\nu_t|\mu) \le e^{-t/\alpha}\mathbf{H}(\nu|\mu), \qquad t > 0 \text{ and } \nu \in \mathbf{M}_1(E),$$

where $\nu_t(\cdot) = \int_E \nu(dx)P_t(x,\cdot)$ and $\mathbf{H}(\nu|\mu) =< f\log(f) >_\mu$, if $d\nu = fd\mu$, $\mathbf{H}(\nu|\mu) = +\infty$ if $\nu \not\ll \mu$, cf. §6.1 of [**2**].

In a recent paper [**1**], BAKRY and EMERY gave a criterion for (L-S) based on the two bilinear forms Γ and Γ_2 defined on $\mathcal{A} \times \mathcal{A}$ by

$$2\Gamma(f,g) = L(fg) - fLg - gLf,$$
$$2\Gamma_2(f,g) = L\Gamma(f,g) - \Gamma(Lf,g) - \Gamma(f,Lg).$$

They showed that, if there is an $\epsilon \in (0,\infty)$ such that

$$\text{(B-E)} \qquad \Gamma_2(f,f) \ge \frac{\epsilon}{2}\Gamma(f,f), \qquad f \in \mathcal{A},$$

then (L-S) holds with $\alpha \le \frac{1}{\epsilon}$. The aim of this note is to extend their result by replacing the constant $\epsilon > 0$ by a bounded measurable function which is not necessarily positive.

To be more precise, for $U \in \mathcal{B}(E;\mathbf{R})$ let $\{P_t^U : t > 0\}$ be the semigroup on $\mathcal{B}(E;\mathbf{R})$ determined by

$$P_t^U f(x) = P_t f(x) - \int_{(0,t)} [P_{t-s}(U \cdot P_s^U f)](x)\,ds, \qquad \text{for } (t,x) \in (0,\infty) \times E,$$

and denote the corresponding GREEN operator by G^U:

$$G^U f(x) \equiv \int_{(0,\infty)} P_t^U f(x)\,dt \qquad f \in \mathcal{B}(E;[0,\infty)).$$

THEOREM 1. *Let $\rho \in \mathcal{B}(E;\mathbf{R})$ be such that*

$$\text{(B-E')} \qquad \Gamma_2(f,f)(x) \ge \frac{\rho(x)}{2}\Gamma(f,f)(x), \qquad f \in \mathcal{A},\ x \in E,$$

then if $\|G^\rho 1\|_{\mathcal{B}} = \sup_{x\in E} G^\rho 1(x) < \infty$, (L-S) holds with $\alpha \le \|G^\rho 1\|_{\mathcal{B}}$.

PROOF: First note that since L is a local operator, L and Γ satisfy the following transformation rules: for $f,g \in \mathcal{A}$ and $\phi \in C^\infty(\mathbf{R},\mathbf{R})$

$$\text{(2)} \qquad L[\phi \circ f] = \phi' \circ f \cdot Lf + \phi'' \circ f \cdot \Gamma(f,f)$$

$$\text{(3)} \qquad \Gamma(\phi \circ f, g) = \phi' \circ f \cdot \Gamma(f,g),$$

cf. [1]. Moreover partial integration yields

$$-< f\cdot Lg >_\mu = < \Gamma(f,g) >_\mu, \qquad f,g\in\mathcal{A}.$$

Next take a strictly positive $f\in\mathcal{A}$ with $\|f\|_2^2 = 1$ and set $g_t = P_t(f^2)$, $u_t = \Gamma(g_t,\log(g_t))$. Then since

$$-\frac{d}{dt} < g_t\log(g_t) >_\mu = < (1+\log(g_t))Lg_t >_\mu = < \Gamma(g_t,\log(g_t)) >_\mu = < u_t >_\mu,$$

and by the ergodic assumption

$$\lim_{t\to\infty} < g_t\log(g_t) >_\mu = 0,$$

we see that

$$< f^2\log(f^2) >_\mu = -\int_{(0,\infty)} \frac{d}{dt} < g_t\log(g_t) >_\mu \, dt = \int_{(0,\infty)} < u_t >_\mu \, dt.$$

Now (3) implies $u_0 = \Gamma(f^2,\log(f^2)) = 4\Gamma(f,f)$, therefore we will get (L-S) as soon as we show that

$$\int_{(0,\infty)} < u_t >_\mu \, dt \le 4\|G^\rho 1\|_B < u_0 >_\mu . \tag{4}$$

In order to show (4) first note that by (2) and (3)

$$L\log(g_t) = \frac{Lg_t}{g_t} - \frac{\Gamma(g_t,g_t)}{g_t^2} = \frac{d}{dt}\log(g_t) - \Gamma(\log(g_t),\log(g_t)).$$

Next the transformation rule for Γ_2 yields

$$\Gamma_2(g_t,\log(g_t)) = g_t\Gamma_2(\log(g_t),\log(g_t)) + \Gamma\big(g_t,\Gamma(\log(g_t),\log(g_t))\big),$$

cf. [1], and therefore with the preceding

$$\begin{aligned}\frac{d}{dt}\Gamma(g_t,\log(g_t)) &= \Gamma(Lg_t,\log(g_t)) + \Gamma(g_t,L\log(g_t)) + \Gamma\big(g_t,\Gamma(\log(g_t),\log(g_t))\big)\\ &= L\Gamma(g_t,\log(g_t)) - 2\Gamma_2(g_t,\log(g_t)) + \Gamma\big(g_t,\Gamma(\log(g_t),\log(g_t))\big)\\ &= L\Gamma(g_t,\log(g_t)) - 2g_t\Gamma_2(g_t,\log(g_t)).\end{aligned}$$

Now under (B-E'), we see that u_t satisfies the differential inequality

$$\frac{d}{dt}u_t \le Lu_t - \rho u_t,$$

from which we deduce $u_t \le P_t^\rho(u_0)$, and since $\{P_t^\rho : t>0\}$ is μ-symmetric,

$$< u_t >_\mu \le < P_t^\rho 1\cdot u_0 >_\mu .$$

Clearly this implies (4) by the definition of G^ρ. ∎

REMARK 5. By the FEYNMAN-KAC formula we have

$$u_t(x) \le P_t^\rho(u_0)(x) = \int_\Omega \exp\left[\int_{(0,t)} -\rho(X_s(\omega))\,ds\right] u_0(X_t(\omega))\,P_x(d\omega).$$

Therefore if $\rho\ge\epsilon>0$, $\|P_t^\rho 1\|_B \le e^{-\epsilon t}$; from this $\|G^\rho 1\|_B \le \frac{1}{\epsilon}$ and we recover BAKRY and EMERY's original result.

It is clear that $\|G^\rho 1\|_B < \infty$ whenever

$$\Lambda(\rho) \equiv \overline{\lim_{t\to\infty}} \frac{1}{t} \log(\|P_t^\rho 1\|_B) < 0.$$

We want to see what can be said about Λ if the law of

$$\omega \in \Omega \longmapsto \mathbf{L}_t(\omega) = \frac{1}{t}\int_{(0,t)} \delta_{X_s(\omega)}\, ds \in \mathbf{M}_1(E)$$

under P_x satisfies a uniform large deviation principle of the form: for all measurable $S \subseteq \mathbf{M}_1(E)$

$$-\inf_{S^o} I \le \underline{\lim_{t\to\infty}} \frac{1}{t}\log\left[\inf_{x\in E} P_x(\mathbf{L}_t \in S)\right] \le \overline{\lim_{t\to\infty}} \frac{1}{t}\log\left[\sup_{x\in E} P_x(\mathbf{L}_t \in S)\right] \le -\inf_{\overline{S}} I,$$

where, for $d\nu = f^2 d\mu$ with $f \in \mathcal{A}$, the rate function I is given by $I(\nu) =< \Gamma(f,f) >_\mu$, cf. §4.2 of **[2]** .

PROPOSITION 6. *Suppose that the above uniform large deviation principle holds and let the operator L have a spectral gap $C > 0$:*

$$\text{(S-G)} \qquad \|f - < f >_\mu\|_2^2 \le \frac{1}{C} < \Gamma(f,f) >_\mu, \qquad f \in \mathcal{A};$$

then $\Lambda(\rho) < 0$ whenever

$$< \rho >_\mu > \frac{(\|\rho - < \rho >_\mu\|_B + \|\rho - < \rho >_\mu\|_2)^2}{4C}.$$

PROOF: From the above large deviation principle we know by VARADHAN's Lemma that

$$\Lambda(\rho) = \sup_{\nu \in \mathbf{M}_1(E)} \left\{ -\int_E \rho\, d\nu - I(\nu)\right\}.$$

Thus the proof will be complete once

$$\text{(7)} \qquad -\int_E \rho\, d\nu + < \rho >_\mu \le \frac{\|\rho - < \rho >_\mu\|_B + \|\rho - < \rho >_\mu\|_2}{C^{1/2}} I(\nu)^{1/2},$$

is verified. In (7) we may assume that $d\nu = f^2 d\mu$ with $f \in \mathcal{A}$. Then using (S-G)

$$\begin{aligned}
-\int_E \rho\, d\nu + < \rho >_\mu &= -\int_E (\rho - < \rho >_\mu)(f^2 - < f >_\mu^2)\, d\mu \\
&= -\int_E (\rho - < \rho >_\mu)(f + < f >_\mu)(f - < f >_\mu)\, d\mu \\
&\le (\|\rho - < \rho >_\mu\|_B \|f\|_2 + \|\rho - < \rho >_\mu\|_2 < f >_\mu)\|f - < f >_\mu\|_2 \\
&\le (\|\rho - < \rho >_\mu\|_B + \|\rho - < \rho >_\mu\|_2) \frac{< \Gamma(f,f) >_\mu^{1/2}}{C^{1/2}}. \qquad \blacksquare
\end{aligned}$$

Finally we want to discuss how the above theory can be applied in the context of symmetric diffusions on a finite dimensional manifold. Throughout E will be an N-dimensional connected complete C^∞ RIEMANNIAN manifold . We will use $(\ |\ ), \lambda, \nabla$ and Δ to denote the inner-product, RIEMANNIAN measure, gradient and LAPLACIAN. Given $V \in C^\infty(E;\mathbf{R})$ such that $Z(V) = \int_E e^{-V}\, d\lambda < \infty$, consider the operator L on $C^\infty(E;\mathbf{R})$ given by

$$Lf \equiv \Delta f - (\nabla V|\nabla f).$$

Then by GAFFNEY's result [**4**], L is an essentially self-adjoint operator on $L^2(\mu)$ where

$$\mu(dx) = \frac{e^{-V(x)}}{Z(V)}\lambda(dx),$$

and there is no problem in constructing the symmetric diffusion process in this setting, cf. Theorem 6.2.9 of [**2**]. Moreover using the BOCHNER-WEITZENBÖCK formula, the operators Γ and Γ_2 are easily identified as

$$\Gamma(f,g) = (\nabla f|\nabla g)$$
$$\Gamma_2(f,f) = \|\mathrm{Hess} f\|^2 + (\mathrm{Ric} + \mathrm{Hess}V)(\nabla f, \nabla f), \qquad f,g \in C^\infty(E;\mathbf{R}),$$

where Ric is the RICCI curvature tensor and Hess$f$ the HESSIAN of f, cf. [**1**]. However unless E is compact, it is not clear how to find the appropriate algebra $\mathcal{A}$. Instead we will introduce the set

$$\mathcal{F} = \left\{f \in \mathcal{B}(E;\mathbf{R}) \cup C^\infty(E;\mathbf{R}) : [Lf] \in L^2(\mu)\right\},$$

which is $\{P_t : t > 0\}$ invariant and for which

$$- < g \cdot Lf >_\mu = < (\nabla f|\nabla g) >_\mu, \qquad f,g \in \mathcal{F},$$

cf. Lemma 6.2.17 of [**2**]. Next let $\rho/2 \in \mathcal{B}(E;\mathbf{R})$ be a lower bound for the symmetric tensor Ric + Hess$V$:

$$\text{(B-E')} \qquad (\mathrm{Ric} + \mathrm{Hess}V)(\nabla f, \nabla f)(x) \geq \frac{\rho(x)}{2}(\nabla f, \nabla f)(x), \qquad x \in E, f \in \mathcal{F}.$$

Then essentially by the same argument of Theorem 1 applied to Lemma 6.2.39 and Theorem 6.2.42 of [**2**], we see that $\|G^\rho 1\|_{\mathcal{B}} < \infty$ implies (L-S) with $\alpha \leq \|G^\rho 1\|_{\mathcal{B}}$.

It should be noted that unlike [**3**], where for compact E estimates of α are obtained in term of the spectral gap of L and lower bounds of ρ, the present argument makes no use of the positiveness of $\|\mathrm{Hess} f\|^2$ in Γ_2.

References

1. D. Bakry and M. Emery, *Diffusions hypercontractives*, in "Séminaire de Probabilités XIX," Springer Lecture Notes in Mathematics **1123**, 1985, pp. 179–206.

2. J.D. Deuschel and D.W. Stroock, "Large Deviations," Academic Press, Boston, 1989.

3. J.D. Deuschel and D.W. Stroock, *Hypercontractivity and spectral gap of symmetric diffusions with applications to the stochastic Ising model*, J. Funct. Ana. (1989) (to appear).

4. M.P. Gaffney, *The conservation property of the heat equation on Riemannian manifolds*, Comm. Pure Appl. Math. **12** (1959), 1–11.

5. L. Gross, *Logarithmic Sobolev inequalities*, Amer. J. Math. **97** (1976), 1061–1083.

Jean-Dominique Deuschel
Department of Mathematics
White Hall
Cornell University
Ithaca, N.Y. 14853

RESCALING THE VACANCY OF A BOOLEAN COVERAGE PROCESS

by

STEVEN N. EVANS[1]

In his recent book [H], Peter Hall gives an encyclopaedic account of the theory of the class of random sets known as *Boolean coverage processes*. We will define this class rigorously in §2, but for the moment we give an intuitive description. Let Π be a homogeneous Poisson point process on $\mathbb{R}^d$ which we enumerate as $\Pi = \{\xi_i\}_{i=1}^{\infty}$. Let $\{S_i\}_{i=1}^{\infty}$ be an independent sequence of independent, identically distributed, random open sets. The Boolean coverage process constructed from the collection of *centres* or *germs*, $\{\xi_i\}$, and the collection of *shapes* or *grains*, $\{S_i\}$, is the random open set $U = \bigcup_i(\xi_i+S_i)$.

Given the random set U, we define the corresponding *vacancy* to be the random measure V given by $V(dx) = 1_{\mathbb{R}^d\setminus U}(x)dx$. In §3.4 of [H], Hall proves strong law of large numbers and central limit theorem type results for the asymptotic vacancy of large sets by a relatively intricate succession of decompositions

---

[1]Research carried out at the University of Virginia and supported in part by NSF Grant DMS-8701212.

and approximations which reduce the problem to one involving arrays of independent, identically distributed, real random variables.

Our aim in this paper is first to prove in §2 that V is an associated random measure (see §1 for a discussion of association), and then to show in §§3 and 4 that Hall's results are simple, direct consequences of the theory developed in [E] for general associated random measures. Although our limit theorems generalise those in [H] to some extent, the main purpose of this paper is to demonstrate how useful and powerful association techniques can be when they are applicable.

We do, however, offer something new by making the observation that, regardless of the common distribution of the shapes $\{S_i\}$, the stationary random measure V is always ergodic.

1. Association and Random Measures

We begin by recalling the notion of an associated random variable (see, for example, [L]). Suppose that $\mathcal{X}$ and $\mathcal{Y}$ are partially ordered sets with orders $\leq_{\mathcal{X}}$ and $\leq_{\mathcal{Y}}$. We say that a map $f: \mathcal{X} \to \mathcal{Y}$ is *non-decreasing* (respectively, *non-increasing*) if $x_1 \leq_{\mathcal{X}} x_2$ implies $f(x_1) \leq_{\mathcal{Y}} f(x_2)$ (respectively, $f(x_1) \geq_{\mathcal{Y}} f(x_2)$). If the set $\mathcal{X}$ has some σ-field of subsets defined on it, and X is an $\mathcal{X}$-valued random variable, then we say that X is *associated* if $\mathrm{Cov}(f(X),g(X)) \geq 0$ for each pair of bounded, measurable, non-decreasing functions $f: \mathfrak{X} \to \mathbb{R}$ and $g: \mathfrak{X} \to \mathbb{R}$ (where we give $\mathbb{R}$ the usual order).

We will be most interested in associated random measures. If Σ is a locally compact metric space, let $M(\Sigma)$ denote the space of Radon measures on Σ topologised by vague convergence and let $\mathcal{M}(\Sigma)$ denote the corresponding Borel σ-field. A *random measure* is a $M(\Sigma)$-valued random variable.

When we speak of association for random measures, we are using the order on $M(\Sigma)$ given by declaring that $\mu \leq \nu$ if $\mu(A) \leq \nu(A)$ for all Borel sets A.

Let $M_p(\Sigma)$ denote the subspace of $M(\Sigma)$ consisting of integer-valued measures; that is, measures μ such that $\mu(A) \in \mathbb{N} \cup \{0\}$ for all Borel sets A. Note that $M_p(\Sigma)$ is a closed subset of $M(\Sigma)$ when Σ is compact; and so in this case $M_p(\Sigma)$ is a locally compact metric space.

2. The Boolean Coverage Process

In order to define Boolean coverage processes rigorously, we must first define what we mean by a random open subset of $\mathbb{R}^d$. We essentially follow §3.1 of [H], but work from a different perspective that will be useful later on.

Suppose that C is a compact subset of $\mathbb{R}^d$. Adjoin an isolated point, Δ, to C and metrise $C \cup \{\Delta\}$ with the metric d_C, where

$$d_C(x,y) = |x-y|, \quad x \neq \Delta \neq y,$$

$$d_C(x,\Delta) = d_C(\Delta,x) = \sup\{|y-z| : y,z \in C\}, \quad x \neq \Delta,$$

$$d_C(\Delta,\Delta) = 0,$$

(here $|\cdot|$ denotes the usual Euclidean distance). It is well known that we can metrise the non-empty, compact subsets of $C \cup \{\Delta\}$ with the Hausdorff metric, ρ_C, defined by

$$\rho_C(J,K) = \sup_{x\in J}\inf_{y\in K} d_C(x,y)$$

$$\vee \sup_{y\in K}\inf_{x\in J} d_C(x,y),$$

and the resulting metric space is compact (see, for example, [D]). It is easy to see that the set $\mathcal{K}_C^\Delta = \{K \subset C \cup \{\Delta\}: K \text{ compact}, \Delta \in K\}$ is a closed, and hence compact, subset of this space. We have a bijection between $\mathcal{K}_C^\Delta$ and the set $\mathcal{K}_C = \{K \subset C: K \text{ compact}\}$ given by $K \longleftrightarrow K \cup \{\Delta\}$. Using this bijection, we can metrise $\mathcal{K}_C$ as a compact metric space.

Now let $\mathcal{O}_C$ denote the set of (relatively) open subsets of C. We have a bijection between $\mathcal{O}_C$ and $\mathcal{K}_C$ given by $G \longleftrightarrow C\backslash G$. Using this bijection, we can therefore also metrise $\mathcal{O}_C$ as a compact metric space.

Let $\mathcal{O}$ denote the set of open subsets of $\mathbb{R}^d$. Define a map $\pi_C: \mathcal{O} \to \mathcal{O}_C$ by $\pi_C(G) = G \cap C$. We define a σ-field, $\mathcal{G}^*$, of subsets of $\mathcal{O}$ to be σ-field generated by the maps π_C, as C ranges over all the compact subsets of $\mathbb{R}^d$ (here, of course, we give $\mathcal{O}_C$ its Borel σ-field).

A random open set, S, is measurable mapping from some underlying probability space $(\Omega,\mathcal{F},P)$ to the measurable space $(\mathcal{O},\mathcal{G}^*)$. We remark that it is possible to show that our definition of $\mathcal{G}^*$ is equivalent to the definition given in §3.1 of [H], and so our definitions of random open set coincide.

Suppose now that $\mathcal{P}$ is a Poisson point process on $\mathbb{R}^d\times\mathcal{O}$ with intensity $\lambda(m\times\sigma)$, where $\lambda \geq 0$, m is Lebesgue measure on $\mathbb{R}^d$ and σ is a probability measure on $(\mathcal{O},\mathcal{G}^*)$. The *Boolean coverage process with characteristics* (λ,σ) is the random open set

$$U = \bigcup_{(\xi,S)\in\mathcal{P}}(\xi+S).$$

The *vacancy* corresponding to U is the random measure, V, given by $V(dx) = 1_{\mathbb{R}^d\setminus U}(x)\ m(dx)$. One can show that V is indeed a random measure in the sense of §1.

The starting point for our investigation of the vacancy is the following theorem which allows us to apply the results of [E].

THEOREM 2.1. The vacancy, V, of the Boolean coverage process, U, with characteristics (λ,σ) is an associated random measure.

PROOF. Set $I = [-\frac{1}{2}, \frac{1}{2}]^d$. For $N \in \mathbb{N}$, consider the Poisson point process $\mathcal{R}_N = \{\pi_{NI}(\xi+S): (\xi,S) \in \mathcal{P}, \xi \in NI\}$. Let R_N be the Poisson point measure corresponding to $\mathcal{R}_N$; that is, the measure we get by placing a unit mass at each of the points of $\mathcal{R}_N$. We may regard R_N as an infinitely divisible $M_p(\mathcal{O}_{NI})$-valued random variable. Applying Corollary 3.5 of [BW2], we see that R_N is associated. The map $H_N: M_p(\mathcal{O}_{NI}) \to M(\mathbb{R}^d)$ defined by $(H_N(\gamma))(dx) = 1_{\mathbb{R}^d\setminus W(\gamma)}(x)\ m(dx)$, where

$W(\gamma) = \bigcup_{S \in \mathrm{supp}\gamma} S$, is continuous and non-increasing. Therefore, by the argument used in the proof of Theorem 3.2 of [L], $V_N = H_N(R_N)$ is an associated random measure on $\mathbb{R}^d$. Finally, we note that $V_N \to V$ almost surely as $N \to \infty$ and so, by Lemma 2.2 and the Remarks following Lemma 2.3 in [E], the random measure V is also associated. ∎

3. Ergodicity and First-Order Asymptotics

The following result seems to be new.

THEOREM 3.1. Let U be a Boolean coverage process with characteristics (λ, σ). The vacancy, V, of U is an ergodic, stationary random measure.

PROOF. If $\lambda = 0$, then $V = m$ almost surely. If $\lambda > 0$ and $\sigma(m) = \infty$, then, by Theorem 3.1 of [H], $V = 0$ almost surely. We may therefore suppose that $\lambda > 0$ and $\sigma(m) < \infty$.

Applying Theorem 3.3 of [E] and the succeeding Remarks, it is certainly enough to show that for any compact set C and any unit vector $u \in \mathbb{R}^d$, we have

$$\lim_{T\to\infty} \mathrm{Cov}(V(C), V(Tu+C)) = 0. \tag{3.1.1}$$

From equation (3.6) in [H] and calculations similar to those on p. 148 of [H], we have

$\mathrm{Cov}(V(C),V(Tu+C))$

$$\le \lambda\int_C\int_C\int m([x_1-x_2-Tu+S]\cap S)\sigma(dS)\ dx_1dx_2$$

$$\le 2\lambda\int_C\int_C\int m\{y\colon |y| > \tfrac{1}{2}|x_1-x_2-Tu|,\ y\in S\}\ \sigma(dS)dx_1dx_2,$$

and (3.1.1) follows by dominated convergence. ■

Ergodic properties of a model similar to U are discussed in [M]. There the shapes forming the covering are random compact sets rather than random open sets. The author gives a sufficient condition for the ergodicity of the coverage process itself in terms of the moments of the radius of a typical shape in the covering.

We can use Theorem 3.1 to give a streamlined proof of the following generalisation of Theorem 3.6 in [H]. For $\mu\in M(\mathbb{R}^d)$ and $T > 0$, we use the notation μ_T to denote the measure defined by $\mu_T(A) = \mu(TA)$, $A\in\mathcal{B}(\mathbb{R}^d)$.

THEOREM 3.2. Let U be a Boolean coverage process with characteristics (λ,σ) and corresponding vacancy V. With probability one, we have that for all $f\in L^1(m)$, $T^{-d}V_T(f)\to\exp\{-\lambda\sigma(m)\}m(f)$ as $T\to\infty$, where we interpret $0\cdot\infty = 0$ and $e^{-\infty} = 0$.

PROOF. For $n\in\mathbb{N}$ and $(k^1,\cdots,k^d)\in\mathbb{N}^d$, set $I(n;\underset{\sim}{k}) = \prod_{i=1}^d[0,k^i2^{-d}[$. Note that if $N\in\mathbb{N}$, then

$$V_N(I(n;\underset{\sim}{k})) = \sum_{r_1=0}^{N-1}\cdots\sum_{r_d=0}^{N-1} V((r_1k^1,\cdots,r_dk^d)2^{-d}+I(n;\underset{\sim}{k})).$$

As $V(x+I(n;\underset{\sim}{k})) \leq m(I(n;\underset{\sim}{k}))$ for any $x \in \mathbb{R}^d$, we see from Theorem VIII.6.9 in [DS] and Theorem 3.1 that $N^{-d}V_N(I(n;\underset{\sim}{k})) \to EV(I(n;\underset{\sim}{k}))$ almost surely as $N \to \infty$ in $\mathbb{N}$. Applying the first moment calculation in §3.2 of [H], it follows easily that

$$(3.2.1) \quad T^{-d}V_T(I(n;\underset{\sim}{k})) \to m(I(n;\underset{\sim}{k}))\exp\{-\lambda\sigma(m)\}$$

almost surely as $T \to \infty$ in $\mathbb{R}$.

For $n \in \mathbb{N}$ and $(k^1,\cdots,k^d) \in \mathbb{Z}^d$, set $J(n;\underset{\sim}{k}) = \prod_{i=1}^d [k^i 2^{-d},(k^i+1)2^{-d}[$. Elementary inclusion-exclusion arguments applied to (3.2.1) show that

$$(3.2.2) \quad T^{-d}V_T(J(n;\underset{\sim}{k})) \to m(J(n;\underset{\sim}{k}))\exp\{-\lambda\sigma(m)\}$$

almost surely as $T \to \infty$ when $k^1,\cdots,k^d \geq 0$; and by similar results for the other orthants, we see that (3.2.2) holds for all $(k^1,\cdots,k^d)$.

Let $\mathcal{C}$ denote the countable class of functions of the form $\Sigma_{\ell=1}^L c_\ell 1_{J(n;\underset{\sim}{k}_\ell)}$ for some $L, n \in \mathbb{N}$, $c_1,\cdots,c_L \in \mathbb{Q}$ and $\underset{\sim}{k}_1,\cdots,\underset{\sim}{k}_L \in \mathbb{Z}^d$. We have shown that, with probability one, $T^{-d}V_T(f) \to \exp\{-\lambda\sigma(m)\}m(f)$ for all $f \in \mathcal{C}$. Since $\mathcal{C}$ is dense in $L^1(m)$, and $T^{-d}V_T \leq m$ for all $T > 0$, the theorem follows. ∎

Hall obtains a less general version of Theorem 3.2 as Theorem 3.6 of [H] using somewhat different methods. Hall shows that $T^{-d}V_T(f) \to \exp\{-\lambda\sigma(m)\}m(f)$ almost surely whenever f is the indicator function of a *Riemann measurable* set (see §3.1 of [H] for a definition of Riemann measurability-- any Riemann measurable set is certainly Lebesgue measurable).

Theorem 3.2 has the following obvious consequence.

COROLLARY 3.3. Under the conditions of Theorem 3.2, $T^{-d}V_T \to \exp\{-\lambda\sigma(m)\}m$ almost surely as $T \to \infty$.

4. Second-Order Asymptotics

THEOREM 4.1. Let U be a Boolean coverage process with characteristics (λ,σ) and corresponding vacancy V. Assume that $\sigma(m^2) < \infty$. Fix $f \in L^1(m) \cap L^\infty(m)$. As $T \to \infty$, $T^{-d/2}(V_T(f)-EV_T(F))$ converges in distribution to a $N(0,\Gamma\|f\|_2^2)$ random variable, where

$$\Gamma = \exp\{-2\lambda\sigma(m)\}\int_{\mathbb{R}^d}(\exp\{\lambda\int m([x+S] \cap S)\sigma(dS)\}-1)dx < \infty.$$

PROOF. Put $I = [-\frac{1}{2}, \frac{1}{2}]^d$. Given Theorem 2.1, the result will follow immediately from Theorem 4.4 in [E] (which generalises Theorem 4.1 in [BW1]) once we show that

$$\sup_{T>0}\mathrm{Cov}(V(I),V(TI)) = \Gamma < \infty.$$

Applying equation (3.6) of [H], we see that

$$\mathrm{Cov}(V(I),V(TI))$$
$$= \exp\{-2\lambda\sigma(m)\}\int_I\int_{TI}(\exp\{\lambda\int m([x_1-x_2+S] \cap S)\sigma(dS)\}-1)\, dx_1dx_2.$$

The quantity on the right hand side increases to Γ as $T \to \infty$ and Theorem 3.4 of [H] gives that $\Gamma < \infty$. ■

Hall obtains Theorem 4.1 as a "rephrasing" of Theorem 3.5 in [H]. The proof of the latter theorem is via an elaborate series of steps which reduce the problem to one for which the classical theory of independent, identically distributed, real random variables applies. Moreover, the results in [H] only cover the special case of the function f being the indicator function of a Riemann measurable set. We also note, without proof, that with a little work it is possible to reverse the order in [H] and obtain an analogous generalisation of Theorem 3.5 in [H] from our Theorem 4.1.

Using Corollary 4.5 in [E], we can obtain a "functional" corollary of Theorem 4.1. We let $S(\mathbb{R}^d)$ denote the usual Schwartz space of rapidly decreasing functions on $\mathbb{R}^d$ and let $S'(\mathbb{R}^d)$ denote the corresponding dual space of tempered distributions.

COROLLARY 4.2. Under the conditions of Theorem 4.1, the map $f \longmapsto (V_T(f)-EV_T(f))$, $f \in S(\mathbb{R}^d)$, defines an $S'(\mathbb{R}^d)$-valued random variable. As $T \to \infty$, $T^{-d/2}(V_T-EV_T)$ converges in distribution to a Gaussian $S'(\mathbb{R}^d)$-valued random variable W with mean 0 and variance given by $EW(f)^2 = \Gamma\int f^2(x)dx$, $f \in S(\mathbb{R}^d)$.

References

[BW1] Burton, R. and Waymire, E. (1985). Scaling limits for associated random measures. *Ann. Probab.* **13**, 1267-1278.

[BW2] Burton, R. and Waymire, E. (1986). The central limit problem for infinitely divisible random measures. In *Dependence in Probability and Statistics* (M. Taqqu, E. Eberlein, eds.). Birkhäuser, Boston.

[D] Debreu, G. (1966). Integration of correspondences. In *Proc. Fifth Berkeley Symp. Math. Statist. and Probab.* **2**, 351-372. University of California Press, Berkeley.

[DS] Dunford, N. and Schwartz, J. T. (1958). *Linear Operators. Part I: General Theory*. Interscience, New York.

[E] Evans, S. N. (1989). Association and random measures. Preprint.

[H] Hall, P. G. (1988). *Introduction to the Theory of Coverage Processes*. Wiley, New York.

[L] Lindqvist, B. H. (1988). Association of probability measures on partially ordered spaces. *J. Multivar. Anal.* **26**, 111-132.

[M] Mase, S. (1982). Asymptotic properties of stereological estimators of volume fraction for stationary random sets. *J. Appl. Probab.* **19**, 111-126.

Department of Statistics
University of California
367 Evans Hall
Berkeley, California 94720

THE BLUMENTHAL-GETOOR-McKEAN THEOREM REVISITED

by

P. J. FITZSIMMONS, R. K. GETOOR, and M. J. SHARPE *

1. Introduction

The Blumenthal-Getoor-McKean theorem **[BGM]** (hereafter referred to as BGM) states that if X and $\tilde{X}$ are two Markov processes with the same hitting distributions, then they may be time changed into each other. This is a deliberately loose statement and one needs to specify the precise hypotheses on X and $\tilde{X}$ and also exactly what the conclusion means before it makes mathematical sense. In §V-5 of **[BG]** a precise statement and proof are given when X and $\tilde{X}$ are standard processes as defined in **[BG]**. It is stated in several places in the literature that the proof in **[BG]** carries over to the case in which X and $\tilde{X}$ are right processes. However, a careful reading of that proof reveals that the quasi-left-continuity (qlc) of X and $\tilde{X}$ is used in a crucial manner at two points: the proofs of (V-5.4) and (V-5.20) in **[BG]**. The purpose of this paper is to give a careful proof of BGM for arbitrary right processes X and $\tilde{X}$ as defined in **[S]**.

It should come as no surprise that, given the development of the technology of Markov processes during the past twenty years, our proof is considerably shorter than that in **[BG]**. Nonetheless the main structure of the proof remains the same. First the theorem is proved under appropriate transience assumptions

* The research of all three authors was supported, in part, by NSF Grant DMS87-21347.

on X and $\tilde{X}$. In general the state space may be decomposed into pieces on which both X and $\tilde{X}$ are transient, hence "locally" related by time change. The second part of the proof concerns the piecing together of these local time changes to obtain the global result. In the proof given in **[BG]** each of these parts contains one of the applications of *qlc* noted above.

Walsh **[W]**, using techniques introduced by Chacon and Jamison **[CJ1,2]** proved a version of BGM more general than that found in **[BG]**. To the best of our knowledge his is the only proof in the literature which covers a reasonably broad class of right processes that are not *qlc*. As we shall indicate at the end of §3, the case in which X and $\tilde{X}$ are *transient Borel* right processes is covered by Theorem 3.4 (and Remark 1 following its statement) in **[W]**. See (4.24c).

In **[Gl2]**, Glover made a remarkable advance by showing that if X and $\tilde{X}$ are transient processes with identical hitting *probabilities*, then the conclusion of BGM still holds. (See also **[F1]** and **[Gl3]**.) Glover shows that his apparently weaker hypotheses actually imply those of BGM. Rather than appealing to BGM he uses transience to finish the proof by a more direct argument. However the assertion [**Gl2**, p. 139], that $U1$ is the potential of a continuous additive functional of Y seems to us to require more justification. Indeed this is exactly the point at which the analogous argument in **[BG]** uses *qlc* in a crucial way. (See [**BG**, V-5.4].) Of course, Glover's claim is true and is an immediate consequence of (3.1) in this paper.

In §2 we develop some consequences of the balayage order that are needed in later sections. Some of these properties are of independent interest. For example, (2.5) and the equivalence of (2.2) and (2.3) show without recourse to BGM that the property of being a (bounded) regular potential depends only on the cone of excessive functions. We also give a simple proof of the fact (due to Getoor and Glover **[GG]**) that an excessive measure dominated by a potential is itself a potential. In §3 we state and prove BGM for transient processes. The general case is treated in §4.

2. Applications of the Balayage Order

Let $X = (\Omega, \mathcal{G}, \mathcal{G}_t, X_t, \theta_t, P^x)$ be a right Markov process in the sense of [**S**, §20] with state space $(E, \mathcal{E})$, semigroup (P_t), and resolvent (U^α). We assume that conditions (20.4) and (20.5) in [**S**] are in force and we shall use the notation in [**S**] with one important exception. Namely we let $\mathcal{E}$ denote the Borel σ-algebra for the original topology on E rather than the Borel σ-algebra for the Ray topology of X on E. We remind the reader that E is a separable Radon space, $S^\alpha = S^\alpha(X)$ is the cone of α-excessive functions $(S = S^0)$, $\mathcal{E}^e = \sigma(\bigcup_{\alpha \geq 0} S^\alpha)$, and $\mathcal{E} \subset \mathcal{E}^e$. The semigroup (P_t) need only be subMarkovian and ζ denotes the lifetime of X. Finally $(\mathcal{F}_t)$ is the usual augmentation of the natural filtration of X, so by [**S**, 20.4i], X is also a right process relative to $(\mathcal{F}_t)$.

Throughout this section we assume that X is *transient* in the sense that there exists a universally measurable g on E with $g > 0$ and $Ug < \infty$. It then follows that:

(2.1) *There exists a finely continuous* $q \in \mathcal{E}^e$ *with* $0 < q \leq 1$ *and* $Uq \leq 1$.

See [**G3**], for example. Also, because X is transient, $\mathcal{E}^e = \sigma(S)$.

Let M be the class of all finite (positive) measures on $(E, \mathcal{E})$. Then the *balayage order*, $\dashv$, on M is defined by

$$\mu \dashv \nu \quad \text{if} \quad \mu(f) \leq \nu(f) \quad \text{for all} \quad f \in S, \tag{2.2}$$

and this is equivalent to

$$\mu \dashv \nu \quad \text{if} \quad \mu U \leq \nu U. \tag{2.3}$$

An $h \in S$ is a *regular potential* provided it is everywhere finite and $P_{T_n} h \to P_T h$ whenever (T_n) is an increasing sequence of $(\mathcal{F}_t)$ stopping times with limit T.

This is equivalent to the existence of a continuous additive functional (CAF), A, of X with $h(x) = P^x(A_\infty)$. See §IV-3 of [**BG**] or [**GS2**]. If λ_n and λ are σ-finite measures on E, then $\lambda_n \downarrow \lambda$ means $\lambda_n(B) \downarrow \lambda(B)$ for all $B \in \mathcal{E}$ with $\lambda_1(B) < \infty$.

(2.4) Proposition. *Let (μ_n) be a sequence in M and $\mu \in M$. Then $\mu_n U \downarrow \mu U$ if and only if $\mu_n(h) \downarrow \mu(h)$ for all bounded regular potentials h.*

Proof. Since Uf is a regular potential provided it is finite, the "if" implication is obvious. Conversely suppose $\mu_n U \downarrow \mu U$. Then $\mu_n \dashv \mu_1$ for all n and since $1 \in S$, $\mu_n(1) \le \mu_1(1)$. Let h be a bounded regular potential with $h \le 1$ for convenience and set $h_n = n\int_0^{1/n} P_t h\,dt$. Then $h_n \uparrow h$. Given $\epsilon > 0$, let $B_n = \{h - h_n > \epsilon\} \in \mathcal{E}^e$. According to (IV-3.6) of [**BG**], which is valid for an arbitary right process, $P^x(T_n < \zeta) \to 0$ as $n \to \infty$ for each x, where $T_n = T_{B_n}$ is the hitting time of B_n. Because $P_t h \to 0$ as $t \to \infty$ it follows that $h_n = Ug_n$ where $g_n = n(h - P_{1/n}h)$. Now by hypothesis, as $n \to \infty$

$$\mu_n(h_k) = \mu_n U g_k \downarrow \mu U g_k = \mu(h_k)$$

for each k, and

$$\mu_n(h - h_k) \le \epsilon\mu_n(B_k^c) + \mu_n(h - h_k; B_k).$$

But $\mu_n(B_k^c) \le \mu_1(1)$, and since $P_{B_k}1 = 1$ on B_k and $P_{B_k}1 \in S$,

$$\mu_n(h - h_k; B_k) \le \mu_n P_{B_k} 1 \le \mu_1 P_{B_k} 1.$$

Moreover

$$\mu_1 P_{B_k} 1 = P^{\mu_1}[T_k < \zeta] \to 0 \quad \text{as} \quad k \to \infty.$$

Combining these facts we see that $\mu_n(h - h_k) \to 0$ as $k \to \infty$ uniformly in n. It now follows that $\mu_n(h) \downarrow \mu(h)$ as $n \to \infty$. ∎

Proposition 2.4 enables us to give the following characterization of bounded regular potentials. See (2.10) for an extension to arbitrary regular potentials and to natural potentials.

(2.5) Proposition. *Let h be a bounded excessive function. Then h is a regular potential if and only if $\mu_n(h) \to \mu(h)$ whenever $(\mu_n) \subset M$ with $\mu_n U \downarrow \mu U$.*

Proof. If h is a regular potential and $\mu_n U \downarrow \mu U$, then (2.4) implies that $\mu_n(h) \to \mu(h)$. Conversely let (T_n) be an increasing sequence of stopping times with limit T. Fix $x \in E$ and let $\mu_n = P_{T_n}(x, \cdot)$ and $\mu = P_T(x, \cdot)$. Clearly $\mu_n U \downarrow \mu U$ and so $P_{T_n} h(x) = \mu_n(h) \to \mu(h) = P_T h(x)$. Thus h is a regular potential. ∎

The next result is due to Meyer [**M2**, p. 417].

(2.6) Proposition. (Meyer). *Let $(\mu_n) \subset M$ and $\lambda \in M$ with $\mu_n \dashv \lambda$ for each n. Then there exist a subsequence (μ_{n_k}) and a $\mu \dashv \lambda$ such that for any bounded potential Ug, $\mu_{n_k} Ug \to \mu Ug$.*

Proposition 2.6 permits a very simple proof of the following result due to Getoor and Glover [**GG**]. Recall that an *excessive measure* ξ is a σ-finite measure on E with $\xi P_t \le \xi$ for each $t \ge 0$. We write Exc for the cone of excessive measures. A measure of the form μU is excessive provided it is σ-finite, which implies that μ itself is σ-finite.

(2.7) Proposition. *Let ξ and λU be excessive with $\xi \le \lambda U$. Then $\xi = \mu U$ for a unique measure μ on E.*

Proof. Since λU is σ-finite we may choose $0 < q \le 1$ with $Uq \le 1$ and $\lambda Uq < \infty$. Now passing to the Uq-transform of X with resolvent $V^\alpha f = (Uq)^{-1} U^\alpha [f \cdot Uq]$, $Uq \cdot \xi$ is excessive for (V^α) and dominated by $(Uq \cdot \lambda) V$. Thus we may suppose λ is finite in proving (2.7). (See pages 298, 299 in [**S**] for the fact that an h-transform of a right process is again a right process.) By XII-38 in [**DM2**] there exists a sequence of finite measures (μ_n) with $\mu_n U \uparrow \xi$. Since $\xi \le \lambda U$, it follows from (2.6) that there exists $\mu \in M$ with $\mu_n U \uparrow \mu U$. Hence $\xi = \mu U$. ∎

(2.8) Remark. A standard reduction now allows one to extend (2.7) to general right processes; that is, (2.7) is valid without assuming that X is transient. See, for example, [**GSt2**, 3.3], [**GSt1**], or [**FM**].

The next fact will be used in §3.

(2.9) Proposition. *Let* τ *be a second countable metrizable topology on* E *such that the Borel* σ*-algebra of* τ, $\mathcal{E}_\tau$, *is contained in* $\mathcal{E}^e$. *Suppose that* $\mu_n(f) \to \mu(f)$ *for every bounded* τ*-continuous function* f *whenever* $\mu_n U \uparrow \mu U$ *for* $\mu \in M$. *Then a.s.* $t \to X_t$ *is* τ*-right continuous.*

Proof. Let f be a bounded τ-continuous function. Under the hypothesis on $\mathcal{E}_\tau$, the process $f \circ X_t$ is nearly optional. See [**S**, p. 21]. Let (T_n) be a sequence of stopping times decreasing to T. Fix $x \in E$ and define μ_n and μ as in the proof of (2.5). This time $\mu_n U \uparrow \mu U$. Consequently

$$P^x[f \circ X_{T_n}; T_n < \zeta] = \mu_n(f) \to \mu(f) = P^x[f \circ X_T; T < \zeta],$$

and so $t \to f \circ X_t 1_{[0,\zeta[}(t)$ is a.s. right continuous. See VI-48 of [**DM1**]. Since τ has a countable base and is metrizable we obtain the conclusion of (2.9). ∎

(2.10) Remark. Recall that a finite excessive function h is a *class* (D) *potential* provided $P_{\{h>n\}}h \to 0$ as $n \to \infty$. See, for example, [**BG**, IV-5.3]. An easy modification of the proof of Proposition 7 in [**R**] shows that a class (D) potential may be written as $\sum h_n$ where each h_n is a bounded excessive function. It now follows from (2.5) that a finite excessive function h is a regular potential if and only if $\mu_n(h) \to \mu(h)$ whenever $(\mu_n) \subset M$, $\mu_1(h) < \infty$, and $\mu_n U \downarrow \mu U$. (For the "if" assertion note first that the condition implies that h is a class (D) potential.) There is a similar characterization of natural potentials. (See [**BG**] or [**GS2**] for the relevant definition.) Namely, a finite excessive function h is a natural potential if and only if $\mu_n(h) \to 0$ for every sequence $(\mu_n) \subset M$ with $\mu_1(h) < \infty$ and $\mu_n U \downarrow 0$.

3. The Transient Case

We begin with the following theorem which shows that for a transient process the cone of excessive functions determines the (bounded) regular potentials.

(3.1) Theorem. *Let X and $\tilde{X}$ be two transient right processes with the same state space E and with $S(X) = S(\tilde{X})$. Then h is a bounded regular potential for X if and only if it is a bounded regular potential for $\tilde{X}$.*

Proof. Letters with tildes denote objects defined over $\tilde{X}$; for example $(\tilde{U}^\alpha)$ is the resolvent of $\tilde{X}$. We emphasize that X and $\tilde{X}$ may be defined on different sample spaces Ω and $\tilde{\Omega}$. In view of (2.5) we need only show that if $(\mu_n) \subset M$, then $\mu_n U \downarrow \mu U$ if and only if $\mu_n \tilde{U} \downarrow \mu \tilde{U}$. Suppose $\mu_n U \downarrow \mu U$. Because $S = \tilde{S}$, the equivalence of (2.2) and (2.3) implies that the sequence of measures $(\mu_n \tilde{U})$ is decreasing and $\mu_n \tilde{U} \geq \mu \tilde{U}$ for each n. Since μ_1 is finite and $\tilde{X}$ is transient, $\mu_1 \tilde{U}$ is σ-finite and hence in $\mathrm{Exc}(\tilde{X})$. Let $\xi = \downarrow \lim \mu_n \tilde{U}$. Then $\xi \in \mathrm{Exc}(\tilde{X})$ and $\mu_1 \tilde{U} \geq \xi \geq \mu \tilde{U}$. By (2.7), $\xi = \nu \tilde{U}$ and $\mu_n \tilde{U} \geq \nu \tilde{U} \geq \mu \tilde{U}$. Using the balayage order again we see that $\mu_n U \geq \nu U \geq \mu U$, and so $\nu U = \mu U$. Hence $\nu = \mu$ by the uniqueness of charges, [**GG**, 1.1]. Therefore $\mu_n \tilde{U} \downarrow \mu \tilde{U}$. Interchanging the roles of X and $\tilde{X}$ completes the proof. ∎

(3.2) Remark. Using (2.10) one sees that X and $\tilde{X}$ have the same regular potentials and the same natural potentials under the hypotheses of (3.1).

Let X be a right process with state space E. For $B \in \mathcal{E}$ and $f \in b\mathcal{E}^u$, let

$$P_B f(x) = P^x[f \circ X_{T_B}; T_B < \infty] \tag{3.3}$$

where $T_B = \inf\{t > 0 : X_t \in B\}$ is the hitting time of B. It is important to note that P_B depends only on the semigroup (P_t) of X; that is, P_B is the same no matter what realization of (P_t) as a right process on E is used in calculating the right hand side of (3.3). See (19.6) and the proof of (19.7) in [**S**]. The next proposition will be used in §4 as well.

(3.4) Proposition. *Let X and $\tilde{X}$ be right processes with the same state space $(E,\mathcal{E})$. Suppose that $P_K(x,\cdot)=\tilde{P}_K(x,\cdot)$ for each $x\in E$ and compact $K\subset E$. Then $S=\tilde{S}$. If X and $\tilde{X}$ are transient, then the equality $S=\tilde{S}$ implies that $P_B(x,\cdot)=\tilde{P}_B(x,\cdot)$ for all $x\in E$ and $B\in\mathcal{E}$.*

Proof. The first assertion is an immediate consequence of Dynkin's theorem [**BG**, II-5.1] or [**S**, 10.32], and [**S**, 10.33]. The converse for transient processes is proved on page 553 of [**F1**]. ■

In this paper a continuous additive functional (CAF), A, of a right process X is a process (A_t) adapted to $(\mathcal{F}_t)$ such that for every ω, $A_0(\omega)=0$ and $t\to A_t(\omega)$ is continuous, increasing, and constant on $[\zeta(\omega),\infty[$, and such that $A_{t+s}(\omega)=A_t(\omega)+A_s(\theta_t\omega)$ identically in t,s, and ω. In view of the known perfection theorems (see [**S**, 55.19], for example) this entails no loss of generality.

We are now prepared to state BGM for transient processes.

(3.5) Theorem. *Let X and $\tilde{X}$ be transient right processes with state space $(E,\mathcal{E})$ and suppose that $P_B(x,\cdot)=\tilde{P}_B(x,\cdot)$ for each $x\in E$ and $B\in\mathcal{E}$. Then there exists a* CAF, *A, of X such that* a.s. *A is strictly increasing and finite on $[0,\zeta[$ and if $a(t)$ is the inverse of A, then the processes $\hat{X}=(X_{a(t)},P^x)$ and $\tilde{X}$ are equivalent.*

(3.6) Remarks. See [**S**, §65] for the precise definition of the time changed process $\hat{X}$ and for the fact that $\hat{X}$ is a right process. Of course, the statement that $\hat{X}$ and $\tilde{X}$ are equivalent means that they have the same semigroup. In view of (12.15) of [**G2**] and the remarks following its proof, one sees easily that the equality of P_K and $\tilde{P}_K$ for all compact K in E implies that they are equal for all $B\in\mathcal{E}$. We emphasize that in Theorem 3.5 we require that $P_B=\tilde{P}_B$ for Borel (or just compact) subsets of E. This should be contrasted with Theorem 4.1. See the remarks (4.3). Moreover, it is shown in [**F1**], that the identity of the hitting *probabilities*, that is, $P_B1=\tilde{P}_B1$ for all $B\in\mathcal{E}$, already implies

$S = \tilde{S}$. Consequently because of (3.4) one may replace the hypothesis $P_B = \tilde{P}_B$ by either (the apparently weaker) $P_B 1 = \tilde{P}_B 1$ or by $S = \tilde{S}$.

Proof. Let $q(\tilde{q})$ be as in (2.1) relative to $X(\tilde{X})$. Then $C_t = \int_0^t q \circ X_s \, ds$ defines a CAF of X which is strictly increasing on $[0, \zeta[$. Let c denote the inverse of C and $Y_t = X_{c(t)}$. If V is the potential kernel of Y, then $V1 = Uq \leq 1$. Let $\tilde{C}, \tilde{c}, \tilde{Y}$, and $\tilde{V}$ be the similar objects relative to $\tilde{X}$ and $\tilde{q}$. It is clear that Y and $\tilde{Y}$ satisfy the hypotheses of (3.5) and, in addition, $V1 \leq 1$ and $\tilde{V}1 \leq 1$. Suppose that (3.5) has been established under this additional condition. Then there exists a CAF, Γ, of Y such that $\hat{Y}_t = Y_{\gamma(t)}$ is equivalent to $\tilde{Y}$ where γ is the inverse of Γ. By [**S**, 19.6]

$$P^x \left[\int_0^t \tilde{q}^{-1} \circ \hat{Y}_s \, ds < \infty, \; t < \zeta(\hat{Y}) \right] = \tilde{P}^x \left[\int_0^t \tilde{q}^{-1} \circ \tilde{Y}_s \, ds < \infty, \; t < \zeta(\tilde{Y}) \right].$$

But $\zeta(\tilde{Y}) = \tilde{C}_{\tilde{\zeta}} = \tilde{C}_\infty$ and

$$\int_0^t \tilde{q}^{-1} \circ \tilde{Y}_s \, ds = \int_0^{\tilde{c}(t)} ds = \tilde{c}(t) < \infty$$

if $t < \tilde{C}_\infty$. Consequently $\hat{C}_t = \int_0^t \tilde{q}^{-1} \circ \hat{Y}_s \, ds$ is a CAF of $\hat{Y}$ that is strictly increasing and finite on $[0, \zeta(\hat{Y})[$, and using [**S**, 19.6] once again, it is evident that $(\hat{Y}_{\hat{c}(t)})$ and $(\tilde{X}_t)$ are equivalent. Then $A_t = \hat{C}(\Gamma(C_t))$ is a CAF of X that is strictly increasing and finite on $[0, \zeta[$ and time changes X into a process equivalent to $\tilde{X}$. (To see that A is $(\mathcal{F}_t)$ adapted one may use an approximation argument as in [**G1**], for example.)

In view of the preceding discussion we may suppose $U1 \leq 1$ and $\tilde{U}1 \leq 1$ in proving Theorem 3.5. By (3.4) we also know that $S = \tilde{S}$. Under these conditions one may repeat the first part of the proof in [**BG**], replacing (V-5.4) in [**BG**] which uses *qlc* by an appeal to (3.1). See the discussion in the second paragraph on page 239 of [**BG**]. Better and cleaner, one may now use the argument on page 139 of [**Gl2**]. The function $\tilde{U}1$ is a bounded regular potential of $\tilde{X}$ and hence of X by (3.1), and now Glover's argument (from (1.12) to the end of §1 of [**Gl2**]) completes the proof of (3.5). ∎

We can actually sharpen the result in (3.5) slightly. See [**F1**] and [**Gl3**]. Namely suppose that there are two different topologies τ and σ on E such that in both topologies E is a separable Radon space and the two topologies have the same Borel σ-algebras: $\mathcal{E}_\tau = \mathcal{E}_\sigma$. If X, resp. $\tilde{X}$, is a transient right process on (E,τ), resp. (E,σ), and $S(X) = S(\tilde{X})$, then the conclusion of (3.5) still holds. The proof depends on the following fact.

(3.7) Proposition. *Let X be a transient right process. If $\mu_n U \uparrow \mu U$ with $\mu \in M$, then $\mu_n(f) \to \mu(f)$ for all bounded continuous functions f on E.*

We shall use (3.7) to establish the assertion of this paragraph and then prove (3.7). Arguing as in the proof of (3.1) we see that $\mu_n U \uparrow \mu U$ if and only if $\mu_n \tilde{U} \uparrow \mu \tilde{U}$. Combining this with (3.7) and (2.9) it follows that $t \to X_t$ is a.s. σ-right continuous, and consequently X is a right process on (E,σ). We may now apply (3.5) to obtain the desired conclusion.

It remains to prove (3.7). Since $\mu_n U \le \mu U$ we may apply Rost's theorem; see [**Ro**], [**M1**], or [**F3**] (where the use of the Kuznetsov measure may be easily avoided) to obtain randomized stopping times T_n with $\mu_n = \mu P_{T_n}$. That is, there exists a probability space $(\Lambda, \mathcal{G}, Q)$ and $T_n : \Omega \times \Lambda \to [0,\infty]$ such that $\omega \to T_n(\omega,\lambda)$ are stopping times for each λ, $T_n \in \mathcal{F} \otimes \mathcal{G}$, and if $\overline{P}^\mu = P^\mu \times Q$, then

$$\mu_n(f) = \int f \circ X_{T_n(\omega,\lambda)}(\omega) \overline{P}^\mu(d\omega, d\lambda).$$

Let q be as in (2.1), so that $t \to q \circ X_t$ is P^μ a.s. right continuous and strictly positive. Then

$$\overline{P}^\mu \int_0^{T_n} q \circ X_t \, dt = \mu U(q) - \mu_n U(q) \to 0$$

as $n \to \infty$. It follows that $T_n \to 0$ in $\overline{P}^\mu$ probability and so if f is a bounded continuous function

$$\mu_n(f) = \overline{P}^\mu(f \circ X_{T_n}) \to \overline{P}^\mu(f \circ X_0) = \mu(f),$$

completing the proof of (3.7). ∎

We close this section by sketching the use of Walsh's theorem [W] to prove (3.5), at least when X and $\tilde{X}$ are Borel. The appropriate transience assumption is now the existence of *Borel* measurable q and $\tilde{q}$ with $Uq \le 1$ and $\tilde{U}q \le 1$. It follows that the processes Y and $\tilde{Y}$ defined in the first paragraph of the proof of (3.5) are Borel. Hence, as before, we may suppose $U1 \le 1$ and $\tilde{U}1 \le 1$. Recall the Ray cone $\mathcal{R}(X)$ as defined in [S, §17] or [G2, §10]. If $f \in b\mathcal{E}$, then $U^\alpha f = Uf - \alpha U\, U^\alpha f$ and so $U^\alpha f \in S(\tilde{X}) - S(\tilde{X})$. Consequently by [S, 17.5] or [G2, 10.2], $t \to f \circ \tilde{X}_t$ is *rcll* a.s. for each $f \in \mathcal{R}(X)$ and similarly $t \to \tilde{f} \circ X_t$ is *rcll* a.s. for each $\tilde{f} \in \mathcal{R}(\tilde{X})$. Let $\mathcal{R}$ be the min-stable cone generated by $\mathcal{R}(X) \cup \mathcal{R}(\tilde{X})$. Then $\mathcal{R}$ is separable in the uniform metric and if $f \in \mathcal{R}$ both $f \circ X$ and $f \circ \tilde{X}$ are *rcll* a.s. Let $\overline{E}$ be the compactification of E_Δ induced by $\mathcal{R}$. Then a.s. both X and $\tilde{X}$ are right continuous on E_Δ in this topology and have left limits in $\overline{E}$. Since $\mathcal{R} \subset b\mathcal{E}_\Delta$, E_Δ is a Borel subset of $\overline{E}$ in the topology induced by $\mathcal{R}$. Now we may realize X and $\tilde{X}$ (more precisely the semigroups (P_t) and $(\tilde{P}_t)$) on the space of all right continuous functions from $[0,\infty[$ to E_Δ which have left limits in $\overline{E}$ and which are absorbed at Δ. One may now apply Theorem 3.4 of [W] to obtain (3.5) of this paper.

4. The General Case

We now state BGM for general right processes.

(4.1) Theorem. *Let X and $\tilde{X}$ be right processes with the same state space $(E,\mathcal{E})$ and cemetery point Δ. Let $E_\Delta = E \cup \{\Delta\}$ and $\mathcal{E}_\Delta = \mathcal{E} \vee \{\Delta\}$. Suppose that*

$$P_B(x,\cdot) = \tilde{P}_B(x,\cdot) \qquad \textit{for all} \qquad x \in E, B \in \mathcal{E}_\Delta. \tag{4.2}$$

Then there exists a CAF, A, of X that is almost surely strictly increasing and finite on $[0,\zeta[$ such that if $a(t)$ is the inverse of A, then $\hat{X} = (X_{a(t)}, P^x)$ and $\tilde{X}$ are equivalent.

(4.3) Remarks. Of course (4.2) also holds for $x = \Delta$. We emphasize that in contrast to the transient case (Theorem 3.5), it is necessary to assume (4.2) for all $B \in \mathcal{E}_\Delta$ and not just in $\mathcal{E}$. Consider the following example. Let E consist of a single point e. X sits at e forever while $\tilde{X}$ sits at e for an exponential holding time and then dies. Clearly X and $\tilde{X}$ have the same hitting distributions on E, but cannot be time changed into one another. It is easy to see that if (4.2) holds for all $B \in \mathcal{E}$ and if, in addition, $P^{\cdot}(\zeta < \infty) = \tilde{P}^{\cdot}(\tilde{\zeta} < \infty)$, then (4.2) holds for all $B \in \mathcal{E}_\Delta$.

The rest of this section is devoted to the proof of Theorem 4.1. Note that (4.2) implies that X and $\tilde{X}$ have the same set of traps D in E. Of course Δ is also a trap for X and $\tilde{X}$. Let (N_k) be a countable base of open sets for the topology of E. Let $\varphi_k = P^{\cdot}[\exp(-T_{E_\Delta \setminus N_k})]$. Then

$$E \setminus D = \cup_k \{x \in E : 1_{N_k}(x)\varphi_k(x) > 0\},$$

and so $D \in \mathcal{E}^e(X)$. Similarly $D \in \mathcal{E}^e(\tilde{X})$. A standard approximation by an increasing sequence of compact sets shows that (4.2) persists for all $B \subset E_\Delta$ such that $B \cap E \in \mathcal{E}^e(X) \cap \mathcal{E}^e(\tilde{X})$. We now adjoin to E a sequence of points (e_n) as isolated points and we modify X and $\tilde{X}$ so that each point in $D \cup \{\Delta\}$ becomes an exponential holding point with mean one from which the modified processes jump successively to $e_1, e_2, \ldots$ where each e_j is an exponential holding point with mean one. Let $E^* = E_\Delta \cup \{e_1, e_2, \cdots\}$ and denote the modified processes by X^* and $\tilde{X}^*$. Using (4.2) extended to $\mathcal{E}^e(X) \cap \mathcal{E}^e(\tilde{X})$ one easily sees that X^* and $\tilde{X}^*$ have the same hitting distributions for all Borel sets in E^*. Clearly X^* and $\tilde{X}^*$ are right processes on E^* with *infinite* lifetimes and no traps. Suppose Theorem 4.1 holds for X^* and $\tilde{X}^*$ and let A^* be the CAF of X^* whose inverse time changes X^* into $\tilde{X}^*$. It is evident that $A^*_t = t - T^*_{D_\Delta} + A^*(T^*_{D_\Delta})$ if $t \geq T^*_{D_\Delta}$ where $D_\Delta = D \cup \{\Delta\}$ and T^*_B denotes the hitting of B by X^*. Therefore $A_t := A^*_{t \wedge T^*_\Delta}$ defines a CAF of X that time changes X into $\tilde{X}$.

Consequently in proving Theorem 4.1 we may suppose that E has no traps and that

$$P_t 1_E(x) = \tilde{P}_t 1_E(x) = 1 \qquad \text{for} \qquad t \geq 0, x \in E. \tag{4.4}$$

These assumptions are in force for the remainder of this section. This reduction is due to J. Glover [**Gl1**].

As noted in the introduction we show first that X and $\tilde{X}$ are related by time changes locally, and then these local time changes are pieced together. We have already seen that (4.2) holds for all $B \in \mathcal{E}^e(X) \cap \mathcal{E}^e(\tilde{X})$. If $B \in \mathcal{E}^e(X)$ and $x \notin B$, then x is regular for B relative to X if and only if $P_B(x, \cdot) = \epsilon_x$. Consequently a set $B \in \mathcal{E}^e(X) \cap \mathcal{E}^e(\tilde{X})$ is finely open for X if and only if it is finely open for $\tilde{X}$. Thus our use of "finely open" in the following lemma is unambiguous. For the remainder of this section we set

$$\mathcal{E}^e = \mathcal{E}^e(X) \cap \mathcal{E}^e(\tilde{X}).$$

Of course, $\mathcal{E} \subset \mathcal{E}^e$. Once Theorem 4.1 is proved, it is easy to see that $\mathcal{E}^e(X) = \mathcal{E}^e(\tilde{X})$ (cf. the argument on page 308 of [**S**]). But we do not see how to prove this directly from (4.2). We shall use two more conventions in the sequel. Firstly we shall omit the phrase "almost surely" in most places where it is obviously required. Secondly we adopt the convention used in §V-5 of [**BG**] of dropping the tilde inside expectations; for example

$$\tilde{P}^x \int_0^\zeta e^{-\alpha t} f \circ X_t \, dt = \tilde{P}^x \int_0^{\tilde{\zeta}} e^{-\alpha t} f \circ \tilde{X}_t \, dt = \tilde{U}^\alpha f(x).$$

(4.5) Lemma. *Let $G \in \mathcal{E}^e$ be finely open. Then $x \to P^x[\exp(-T_G)]$ and $x \to \tilde{P}^x[\exp(-T_G)]$ are $\mathcal{E}^e$ measurable.*

Proof. Let $T = T_G$ (and $\tilde{T} = \tilde{T}_G$). Since $G \in \mathcal{E}^e$ and G is finely open,

$$H := \{x : P^x(T > 0) = 1\} = \{x : \tilde{P}^x(T > 0) = 1\}.$$

Thus H is the common state space for the processes (X,T) and $(\tilde{X},\tilde{T})$ obtained by killing X (resp. $\tilde{X}$) at T (resp. $\tilde{T}$). Since

$$H = \{x: P^x(e^{-T}) < 1\} = \{x: \tilde{P}^x(e^{-T}) < 1\},$$

it follows that $H \in \mathcal{E}^e$. The Borel σ-algebra $\mathcal{H}$ of H is the trace of $\mathcal{E}$ on H and so $\mathcal{H} \subset \mathcal{E}^e$. It now follows easily from (4.2) (extended to $B \in \mathcal{E}^e$) that (X,T) and $(\tilde{X},\tilde{T})$ have the same hitting distributions for sets in $\mathcal{H}$. See [**BG**, V-5.2]. Therefore by (3.4), $S(X) = S(\tilde{X})$ and $S(X,T) = S(\tilde{X},\tilde{T})$. It follows easily from the identity $V^\alpha f = U^\alpha f - P^\alpha_T U^\alpha f$ for $\alpha > 0$ and $f \in b\mathcal{E}^u$, where (V^α) is the resolvent of (X,T), that $S(\tilde{X},\tilde{T}) = S(X,T) \subset \mathcal{E}^e(X)$. Similarly $S(X,T) \subset \mathcal{E}^e(\tilde{X})$. Now let $\varphi = P^\cdot[\exp(-T)]$. Then $\varphi = P^\cdot(T < \infty) - V^0\varphi$ and by the above remarks $\varphi \in \mathcal{E}^e(\tilde{X})$. Similarly $\tilde{P}^\cdot[\exp(-T)] \in \mathcal{E}^e(X)$. ∎

If $B \in \mathcal{E}^e$ we write $\tau_B = \tau(B) = T_{E\setminus B}$ for the exit time from B, and $B^c = E\setminus B$. Following [**BG**] we say that a set $F \subset E$ is an *exit set* provided $F \in \mathcal{E}^e$, F^c is finely open, and $\tilde{P}^x(\tau_F < \infty) = 1$ for all $x \in E$ (equivalently by (4.2), $P^x(\tau_F < \infty) = 1$ for all $x \in E$). As we shall see E may be covered by a countable collection of exit sets, and our first goal is to relate X and $\tilde{X}$ on each exit set.

Fix an exit F and consider the subprocesses (X,τ_F) and $(\tilde{X},\tilde{\tau}_F)$. The common state space for these processes is

$$E_F := \{x \in E: P^x(\tau_F > 0) = 1\} = \{x \in E: \tilde{P}^x(\tau_F > 0) = 1\},$$

where the equality follows as in the proof of (4.5)— $G := F^c$ is finely open and $\tau_F = T_G$. As in the proof of (4.5), (X,τ_F) and $(\tilde{X},\tilde{\tau}_F)$ have the same hitting distributions for sets in $\mathcal{E}_F$ —the trace of $\mathcal{E}$ on E_F. Since F is an exit set both (X,τ_F) and $(\tilde{X},\tilde{\tau}_F)$ are *transient* right processes on the common state space $(E_F,\mathcal{E}_F)$.

(4.6) Proposition. *Let F be an exit set. Then there exists a* CAF, A^F , *of* (X, τ_F) *which is strictly increasing and finite on* $[0, \tau_F]$ *such that* $(\tilde{X}, \tilde{\tau}_F)$ *is equivalent to* (X, τ_F) *time changed by (the inverse of)* A^F .

Proof. In view of the preceding remarks this is an immediate consequence of (3.5), except for the assertion that $A^F(\tau_F) < \infty$. But the time change of (X, τ_F) by A^F has lifetime $A^F(\tau_F)$ which must be finite since the equivalent process $(\tilde{X}, \tilde{\tau}_F)$ has finite lifetime $\tilde{\tau}_F$. ∎

Fix an exit set F and let A^F be as in (4.6). By an extension procedure detailed in [**S**, 69.3] (see also [**GS1**]) there is a unique diffuse optional perfect homogeneous random measure (HRM), κ^F , of X which is carried by M , the union over positive rational r of the intervals $]r, r + \tau_F \circ \theta_r[$, such that

$$\kappa^F(\omega, [0,t]) = A_t^F(\omega), \qquad \text{for all} \qquad t \in [0, \tau_F(\omega)]. \tag{4.7}$$

Since $\{t: X_t \in F\} \backslash M$ is countable, κ^F is also the unique diffuse optional HRM carried by $\{t: X_t \in F\}$ and satisfying (4.7). Since A^F is strictly increasing up to time τ_F , κ^F charges each nonvoid interval $]r, r+\tau_F \circ \theta_r[$. The set F^c being finely open, if $\tau_F = 0$ then $T_F > 0$ and $M \cap [0, \epsilon]$ is empty for all small $\epsilon > 0$. It now follows from (4.7) that $\lim_{t \downarrow 0} \kappa^F([0,t]) = 0$. Consequently $t \to \kappa^F([0,t])$ is a CAF of X which is finite and strictly increasing on $[0, \tau_F]$. As in [**BG**, V-5.11] we have the following compatibility relationship for any two exit sets F_1 and F_2 :

$$A_t^{F_1} = A_t^{F_1 \cap F_2} = A_t^{F_2} \qquad \text{for all} \qquad t \in [0, \tau_{F_1} \wedge \tau_{F_2}], \quad \text{a.s.} \tag{4.8}$$

(4.9) Lemma. *If F_1 and F_2 are exit sets, then as measures on* $[0, \infty[$,

$$1_{F_1 \cap F_2} \circ X_t \, \kappa^{F_1}(dt) = 1_{F_1 \cap F_2} \circ X_t \, \kappa^{F_2}(dt), \quad \text{a.s.} \tag{4.10}$$

Proof. Both sides of (4.10) are diffuse optional HRM's carried by $\{t: X_t \in F_1 \cap F_2\}$, and they agree on $[0, \tau_{F_1 \cap F_2}]$ because of (4.8). By the uniqueness of the extension procedure described above (as applied to $F_1 \cap F_2$), both sides of (4.10) coincide with $\kappa^{F_1 \cap F_2}$, and the lemma follows. ∎

Fix $x \in E$. Since x is not a trap for $\tilde{X}$ there is an open neighborhood N of x such that $\tilde{P}^x(\tau_{\bar{N}} < \infty) > 0$. (Here $\bar{N}$ is the closure of N.) Arguing as on page 240 of [BG] we see that if $\tilde{\varphi} := \tilde{P}^{\cdot}\,(\exp(-\tau_{\bar{N}}))$ then $F := \bar{N} \cap \{\tilde{\varphi} \geq \epsilon\}$ is an exit set for any $\epsilon > 0$. ($F \in \mathcal{E}^e$ because of (4.5).) Let $\{N_i\}$ be a countable base for the topology of E and put $\tilde{\varphi}_i = \tilde{P}^{\cdot}(\exp(-\tau_{\bar{N}_i}))$. Then $F_{ij} := \bar{N}_i \cap \{\tilde{\varphi}_i \geq 1/j\}$ is a countable collection of exit sets with union E. Moreover E is the union of $N_i \cap \{\tilde{\varphi}_i > 1/j\}$ and so the union of the $E_{F_{ij}}$ as well. Relabel the F_{ij} as $(F_n, n \geq 1)$, define $C_1 = F_1$, $C_{n+1} = F_{n+1} \backslash (F_1 \cup \cdots \cup F_n)$, and set

$$\kappa(dt) = \sum_n 1_{C_n}(X_t)\kappa^{F_n}(dt). \tag{4.11}$$

Clearly κ is an optional perfect HRM of X and it is easy to check that almost surely κ charges every open subinterval of $[0, \infty[$. It follows readily from (4.9) that for any exit set F

$$\kappa([0,t]) = A_t^F \qquad \text{for all} \qquad t \in [0, \tau_F], \ \text{a.s.} \tag{4.12}$$

Define $A_t = \kappa([0,t])$. Because of (4.12), $A_{0+} = 0$ and the argument at the top of page 247 of [BG] shows that $t \to A_t$ is continuous. Therefore A is a CAF of X which is finite and strictly increasing on $[0, R[$ where $R = \inf\{t: A_t = \infty\}$, and $P^x(R > 0) = 1$ for every x.

As noted previously (4.2) holds for all $B \in \mathcal{E}^e$ and in particular for all exit sets. Let Y be the time change of X by the inverse of A. Then Y has lifetime $A(R) = \infty$. It follows from (4.6) that Y killed when it first leaves an exit set F and $(\tilde{X}, \tilde{\tau}_F)$ are equivalent, and hence so are the subprocesses obtained by killing them at the first time they hit $B \in \mathcal{E}^e$ provided $F^c \subset B$. Therefore the next result follows by the argument used to prove (V-5.25) in [BG].

(4.13) Proposition. *Let F be an exit set and let $B \in \mathcal{E}^e$ with $F^c \subset B$. Then*

$$P^x(e^{-\alpha A(T_B)} f \circ X_{T_B}) = \tilde{P}^x(e^{-\alpha T_B} f \circ X_{T_B})$$

for all $x \in E$, $\alpha > 0$, and $f \in p\mathcal{E}^u$.

We shall use (4.13) to prove that A is the CAF promised by Theorem 4.3. First we must remove the constraint $F^c \subset B$ in (4.13). To this end we require two lemmas whose proofs are deferred to the end of this section.

(4.14) Lemma. *Let (F_n) be a sequence of exit sets and define stopping times for X as follows: $S_0 = 0$, $S_{n+1} = S_n + \tau_{F_{n+1}} \circ \theta_{S_n}$, $n \geq 0$. Let $(\tilde{S}_n; n \geq 0)$ denote the analogous sequence for $\tilde{X}$. Then*

$$P^x(e^{-\alpha A(T_B \wedge S_n)} f \circ X_{T_B \wedge S_n}) = \tilde{P}^x(e^{-\alpha(T_B \wedge S_n)} f \circ X_{T_B \wedge S_n}), \tag{4.15}$$

for all $B \in \mathcal{E}^e$, $\alpha > 0$, $x \in E$, and $f \in p\mathcal{E}^u$.

(4.16) Lemma. *There is a sequence (F_n) of exit sets such that if $(\tilde{S}_n; n \geq 0)$ is defined as in (4.14), then (a.s. for $\tilde{X}$), $\tilde{S}_n < \infty$ and $\tilde{S}_n \uparrow \infty$.*

(4.17) Proposition. *For $x \in E$, $\alpha > 0$, and $B \in \mathcal{E}^e$,*

$$P^x(e^{-\alpha A(T_B)}) = \tilde{P}^x(e^{-\alpha T_B}).$$

Proof. Let (F_n) be the sequence of exit sets in (4.16) and let (S_n) and $(\tilde{S}_n)$ be defined as in (4.14). Using (4.15) with B empty and $f = 1$ yields for $\alpha > 0$ and $x \in E$ fixed

$$P^x(e^{-\alpha A(S_n)}) = \tilde{P}^x(e^{-\alpha S_n}). \tag{4.18}$$

Let $R = \inf\{t: A_t = \infty\}$, $S := \uparrow \lim S_n$. By (4.16) and (4.18), $A(S) = \lim A(S_n) = \infty$, and so $S \geq R$. Since $A_t = \infty$ if $t \geq R$, we have $A(T_B \wedge S) = A(T_B)$. Letting $n \to \infty$ in (4.15) with $f = 1$, we obtain (4.17). ■

Because of the remarks below (4.12), $R = \inf\{t: A_t = \infty\}$ is a perfect exact thin terminal time. Note that A is a CAF of (X, R) which is finite and strictly increasing on $[0, R[$.

(4.19) Proposition. *Let $\hat{X}$ denote the time change of (X, R) by A. Then $\hat{X}$ is equivalent to $\tilde{X}$.*

Proof. The lifetime of $\hat{X}$ is $A(R) = \infty$. Moreover, if $B \in \mathcal{E}$ then $\inf\{t: \hat{X}_t \in B\} = A(T_B)$. Letting $\hat{P}^x$ denote the law of $\hat{X}$ starting at x, (4.17) becomes

$$\hat{P}^x(e^{-\alpha T_B}) = \tilde{P}^x(e^{-\alpha T_B}) \tag{4.20}$$

valid for $x \in E$, $\alpha > 0$, and $B \in \mathcal{E}$. The content of (4.20) is that the α-subprocesses of $\hat{X}$ and $\tilde{X}$ have identical hitting probabilities. These α-subprocesses are transient, so by **[F1]**, they have the same hitting distributions and then by (3.5) they are (equivalent to) time changes of each other. But as noted in **[F2]** this fact for just one $\alpha > 0$ implies that $\hat{X}$ and $\tilde{X}$ are equivalent processes. ∎

We now show that $R = \infty$. In view of (4.19) this will complete the proof of Theorem 4.1. Letting $\alpha \to 0$ in (4.20) and using (4.2) and (4.19) we see that all four processes X, $\tilde{X}$, $\hat{X}$, and (X, R) have the same hitting probabilities. It follows that for $x \in E$

$$P^x(T_B < \infty) = P^x(T_B < R), \tag{4.21}$$

first for all $B \in \mathcal{E}$ and then for all $B \in \mathcal{E}^e(X)$. Since the hypotheses on X and $\tilde{X}$ are symmetric we may apply (4.16) to X to obtain a sequence (G_n) of exit sets such that if $U_0 = 0$, $U_{n+1} = U_n + \tau_{G_{n+1}} \circ \theta_{U_n}$, $n \geq 0$, then (a.s. for X), $U_n < \infty$ and $U_n \uparrow \infty$. Suppose that $P^x(R < \infty) > 0$ for some x. Since $R > 0$, there exists an $n \geq 0$ with $P^x(U_n < R \leq U_{n+1}) > 0$. But R is a terminal time, and so

$$0 < P^x(U_n < R \leq U_{n+1}) = P^x[P^{X(U_n)}(R \leq \tau_{G_{n+1}})] = 0,$$

by (4.21) since, G_{n+1} being an exit set, $P^y(\tau_{G_{n+1}} < \infty) = 1$ for all y. Therefore $R = \infty$.

It remains to establish (4.14) and (4.16).

Proof of (4.14). It suffices to prove (4.14) for compact B. Fix such a B, and fix $x \in E$, $\alpha > 0$. Since each F_n^c is finely open, an induction argument using (4.2) and Blumenthal's zero-one law shows that $P^x(S_n = 0) = \tilde{P}^x(S_n = 0)$ for all $n \geq 0$. Let $N = \sup\{n \geq 0 \colon P^x(S_n = 0) = 1\}$. If $N = \infty$, (4.15) is obvious, so we may assume $N < \infty$. Clearly (4.15) holds for $n \leq N$. If $n > N$, then $P^x(S_n > 0) = \tilde{P}^x(S_n > 0) = 1$. In particular, a.s. P^x, $S_{N+1} = \tau(F_{N+1})$ and $T_B \wedge S_{N+1} = T(B \cup F_{N+1}^c)$, and the analogous statement holds for the same quantities relative to $\tilde{X}$. Thus (4.15) reduces to (4.13) when $n = N+1$. We now proceed by induction taking $n = N+1$ as the initial case. Observe that if $n > N$, then $S_n > 0$ a.s. P^x and so

$$(4.22) \qquad T_B > S_n \Longleftrightarrow X_{T_B \wedge S_n} \notin B \qquad \text{a.s. } P^x$$

since B is closed. Of course, the analogous statement holds for $\tilde{X}$. By considering separately the cases $S_{n+1} \leq T_B$ and $S_n < T_B < S_{n+1}$, one checks that $T_B \wedge S_{n+1} = S_n + T(B \cup F_{n+1}^c) \circ \theta_{S_n}$ on $\{S_n < T_B\}$ (and similarly for the same objects with tildes). Now using (4.22) we have

$$\begin{aligned}
(4.23)\, & P^x(\exp[-\alpha A(T_B \wedge S_{n+1})] f \circ X(T_B \wedge S_{n+1})) \\
&= P^x(\exp[-\alpha A(T_B \wedge S_n)](1_B f) \circ X(T_B \wedge S_n)) \\
&\quad + P^x[\exp[-\alpha A(T_B \wedge S_n)](1_{E \setminus B}) \circ X(T_B \wedge S_n) \\
&\qquad \times P^{X(T_B \wedge S_n)}(\exp[-\alpha A(T(B \cup F_{n+1}^c))] f \circ X(T(B \cup F_{n+1}^c)))].
\end{aligned}$$

In view of (4.13) the inner expectation in the second term on the right hand side of (4.23) equals

$$\tilde{P}^{X(T_B \wedge S_n)}[\exp[-\alpha T(B \cup F_{n+1}^c)] f \circ X(T(B \cup F_{n+1}^c))].$$

Invoking the induction hypothesis and reversing the steps in the computation in (4.23) we arrive at the desired equality for $n+1$. This completes the proof of (4.14). ∎

Proof of (4.16). For typographical reasons we shall write the proof for X rather than $\tilde{X}$. Let $(H_i,\ i \geq 1)$ be a sequence of exit sets such that E is the union of the E_{H_i}. Such a sequence was constructed in the paragraph following the proof of (4.9). For $i,j,k \geq 1$ define

$$K(i,j) = \{x: P^x[\exp(-\tau_{H_i})] < 1 - 1/j\}$$

$$\varphi_{ij}(x) = P^x[\exp(-T_{K(i,j)})]$$

$$L(i,j,k) = \{\varphi_{ij} < 1/k\}.$$

Note that $K(i,j) \subset H_i$ and that $\cup_j K(i,j) = E_{H_i}$ so that E is the union of the $K(i,j)$. Moreover $H_i \cap L^c(i,j,k)$ and $K^c(i,j) \cap L^c(i,j,k)$ are exit sets. Construct a sequence (F_n) of exit sets of the above form such that for each triple (i,j,k), $F_n = K^c(i,j) \cap L^c(i,j,k)$ and $F_{n+1} = H_i \cap L^c(i,j,k)$ for *infinitely* many $n \geq 1$. Let S_n be as in (4.16). Since each F_n is an exit set, it follows that $S_n < \infty$ for each n. But X can move from $K(i,j)$ to H_i^c only finitely many times in any finite time interval and so writing $T(i,j,k)$ for the hitting time of $L(i,j,k)$, we must have

$$S := \uparrow \lim S_n \geq T(i,j,k)$$

for each triple (i,j,k). Let $T(i,j) := \uparrow \lim_k T(i,j,k) \leq S$. Since $\varphi_{ij} \circ X_{T(i,j,k)} \leq 1/k$ on $\{T(i,j,k) < \infty\}$ and since $e^{-t}\varphi_{ij} \circ X_t$ is a supermartingale, we have $\varphi_{ij} \circ X_{T(i,j)} = 0$ on $\{T(i,j) < \infty\}$. This in turn implies that $\varphi_{ij} \circ X_S = 0$ on $\{S < \infty\}$. But $\varphi_{ij} = 1$ on $K(i,j)$ and since E is the union over all i,j of the $K(i,j)$, $\sup_{i,j} \varphi_{ij} = 1$. It follows that $S = \infty$ as required, proving (4.16). ∎

(4.24) Remarks. (a) We emphasize that the proof just completed shows that Theorem 4.1 is a result about the right *semigroups* (P_t) and $(\tilde{P}_t)$ and does not depend on the particular realizations X and $\tilde{X}$ of them as right processes.

(b) As in the transient case Theorem 4.1 is valid if X and $\tilde{X}$ are right processes relative to different (separable Radon) topologies τ and σ on E provided the associated Borel σ-algebras coincide. Indeed, it still follows that a set in $\mathcal{E}^e(X) \cap \mathcal{E}^e(\tilde{X})$ is finely open for X if and only if it is finely open for $\tilde{X}$, and Lemma 4.5 remains true as well (although one must use a little more care in deducing the relevant part of Proposition 3.4 from Dynkin's theorem and [**S**, 10.33]). These facts are enough to repeat the proof of (4.16) to obtain a sequence of exit sets (F_n) such that the stopping times $S_0 := 0$, $S_{n+1} := S_n + \tau(F_{n+1}) \circ \theta_{S_n}$ are all finite and $S_n \uparrow \infty$. It follows as in §3 that (X, τ_{F_n}) is a.s. σ-right continuous, so by induction X is a.s. σ-right continuous on each interval $[0, S_n[$, hence on $[0, \infty[$. Thus X is actually a right process on (E, σ) and Theorem 4.1 now applies.

(c) We saw in §3 how to reduce BGM to a theorem of Walsh when X and $\tilde{X}$ are transient Borel right processes. We do not know whether a similar reduction can be made without the transience hypothesis. The problem is to find a Lusin topological space in which both X and $\tilde{X}$ are *rcll* (for Walsh's method makes critical use of the existence of left limits). In the general case we are unable to deduce directly from (4.2) that $t \to U^\alpha f \circ \tilde{X}_t$ and $t \to \tilde{U}^\alpha f \circ X_t$ are a.s. *rcll*. This property is, of course, a consequence of Theorem 4.1.

References

[BG] R. M. Blumenthal and R. K. Getoor. ***Markov Processes and Potential Theory.*** Academic Press, New York, 1968.

[BGM] R. M. Blumenthal, R. K. Getoor, and H. P. McKean, Jr. Markov processes with identical hitting distributions. Ill. J. Math., **6** (1962), 402–420, and supplement Ill. J. Math., **7** (1963), 540–542.

[CJ1] R. V. Chacon and B. Jamison. A fundamental property of Markov processes with an application to equivalence under time changes. Israel J. Math., **33** (1979), 241–269.

[CJ2] R. V. Chacon and B. Jamison. Processes with state dependent hitting probabilities. Adv. in Math., **32** (1979), 1–35.

[DM1] C. Dellacherie et P. A. Meyer. *Probabilités et Potentiel*, II. Hermann, Paris, 1980.

[DM2] C. Dellacherie et P. A. Meyer. *Probabilités et Potentiel*, IV. Hermann, Paris, 1987.

[F1] P. J. Fitzsimmons. Markov processes with identical hitting probabilities. Math. Z., **192** (1986), 547–554.

[F2] P. J. Fitzsimmons. On the identification of Markov processes by the distribution of hitting times. Sem. Stoch. Proc., 1986, 15–19. Birkhäuser, Boston, 1987.

[F3] P. J. Fitzsimmons. Penetration times and Skorohod stopping. Sém. de Prob. XXII, Lecture Notes in Math., **1321**, Springer, Berlin-Heidelberg-New York, 1988.

[FM] P. J. Fitzsimmons and B. Maisonneuve. Excessive measures and Markov processes with random birth and death. Probab. Th. Rel. Fields, **72** (1986), 319–336.

[G1] R. K. Getoor. Some remarks on continuous additive functionals. Ann. Math. Stat., **38** (1967), 1655–1660.

[G2] R. K. Getoor. *Markov Processes: Ray Processes and Right Processes.* Lecture Notes in Math., **440**, Springer, Berlin-Heidelberg-New York, 1975.

[G3] R. K. Getoor. Transience and recurrence of Markov processes. Sém. de Prob. XIV, Lecture Notes in Math., **784**, Springer, Berlin-Heidelberg-New York, 1980.

[GG] R. K. Getoor and J. Glover. Markov processes with identical excessive measures. Math. Z., **184** (1983), 287–300.

[GS1] R. K. Getoor and M. J. Sharpe. Balayage and multiplicative functionals. Z. Wahrscheinlichkeitstheorie verw. Geb., **28** (1974), 139–164.

[GS2] R. K. Getoor and M. J. Sharpe. Naturality, standardness, and weak duality for Markov processes. Z. Wahrscheinlichkeitstheorie verw. Geb., **67** (1984), 1–62.

[GSt1] R. K. Getoor and J. Steffens. The energy functional, balayage, and capacity. Ann. Inst. Henri Poincaré, **23** (1987), 321–357.

[GSt2] R. K. Getoor and J. Steffens. More about capacity and excessive measures. Sem. Stoch. Proc., 1987, 135–157. Birkhäuser, Boston, 1988.

[Gl1] J. Glover. Representing last exit potentials as potentials of measures. Z. Wahrscheinlichkeitstheorie verw. Geb., **61** (1982), 17–30.

[Gl2] J. Glover. Markov processes with identical hitting probabilities. Trans. Amer. Math. Soc., **275** (1983), 131–141.

[Gl3] J. Glover. Identifying Markov processes up to time change. Sem. Stoch. Proc., 1982, 171–194. Birkhäuser, Boston, 1983.

[M1] P. A. Meyer. Le schéma de remplissage en temps continu, d'après H. Rost. Sém. de Prob. VI, Lecture Notes in Math., **258**, Springer, Berlin-Heidelberg-New York, 1972.

[M2] P. A. Meyer. Convergence faible et compacité des temps d'arrêt, d'après Baxter et Chacon. Sém. de Prob. XII, Lecture Notes in Math., **649**, Springer, Berlin-Heidelberg-New York, 1978.

[R] M. Rao. A note on Revuz measure. Sém. de Prob. XIV, Lecture Notes in Math., **784**, Springer, Berlin-Heidelberg-New York, 1980.

[Ro] H. Rost. The stopping distributions of a Markov process. Invent. Math., **14** (1971), 1–16.

[S] M. J. Sharpe. *General Theory of Markov Processes.* Academic Press, San Diego, 1988.

[W] J. B. Walsh. On the Chacon-Jamison theorem. Z. Wahrscheinlichkeitstheorie verw. Geb., **68** (1984), 9–28.

Department of Mathematics, C-012
University of California, San Diego
La Jolla, California 92093

LOCAL TIMES, OCCUPATION TIMES, AND THE LEBESGUE MEASURE OF THE RANGE OF A LEVY PROCESS

by

P. J. FITZSIMMONS* and S. C. PORT

0. Introduction

Let $X = (X_t: t \geq 0)$ be a Lévy process on the line for which singletons are non-polar. We assume that X *does not* have the form $\tilde{X}_t + bt$, where $\tilde{X}$ is a compound Poisson process. Let $N_t(a)$ denote the occupation time, up to time t, of $(0, a]$ if $a > 0$, and the negative of the occupation time, up to time t, of $(a, 0]$ if $a \leq 0$. Let $R_t(a)$ denote the Lebesgue measure of the partial range $\{X_s: 0 \leq s \leq t\}$ intersected with $(0, a]$ if $a > 0$ (and the negative of the measure of this range intersected with $(a, 0]$ if $a \leq 0$). Our purpose in this paper is to investigate the L^2-differentiability of $a \mapsto N_t(a)$ and $a \mapsto R_t(a)$. As it turns out, the derivatives of $N_t(a)$ coincide with certain "local times," even when singletons are semipolar for X. These local times also arise as limits of upcrossing and downcrossing processes. In the following discussion we consider the cases "0 regular for $\{0\}$" and "0 irregular for $\{0\}$" separately.

If 0 is regular (for $\{0\}$), then x is regular for $\{x\}$, for each $x \in \mathbf{R}$. In this case X satisfies hypothesis (F) of Hunt [8], and relative to Lebesgue measure X has a λ-potential kernel density $g^\lambda(x, y)$ that is continuous in (x, y). There is a large

* Research supported in part by NSF grant DMS 8419377.

literature on characterizations of local times for such processes; see, e.g., [1], [2], [3], [5], [6], [7], and the references therein. Local time as derivative of occupation time was first investigated, for Hunt processes satisfying (F), by Griego [7]. More recently, Bally [1] has obtained definitive results concerning L^2-convergence of additive functionals, in case 0 is regular. For this reason we exclude the regular case except in our study of $R_t(a)$. However, obvious modifications of the proof of our Theorem 1 yield the analogous result in the regular case.

Suppose now that 0 is irregular for $\{0\}$. In this case results from [4] imply that

$$\log E_0(e^{i\theta X_1}) = i\theta b + \int (e^{i\theta x} - 1)\nu(dx),$$

where $b \neq 0$, $\nu(\mathbf{R}\backslash\{0\}) = \infty$, $\int(|x| \wedge 1)\, \nu(dx) < \infty$. Since the Lévy measure ν has infinite mass, the law of X_t has no atoms whenever $t > 0$. For each a the visiting set $\{s \in (0,t]: X_s = a\}$ is finite for all $t > 0$ (almost surely), and the counting process $t \mapsto \mathrm{card}\{s \in (0,t]: X_s = a\}$ is a delayed renewal process. For fixed $t > 0$ and $a \in \mathbf{R}$, $x \to N_t(x)$ is L^2-differentiable at a from the right and from the left. If $b < 0$ then the right derivative is just $\mathrm{card}\{s \in (0,t]: X_s = a\}/|b|$, while the left derivative is $\mathrm{card}\{s \in [0,t): X_s = a\} \,/\, |b|$. (A similar result holds if $b > 0$.) Moreover, if $P(X_0 = a) = 0$, then the left and right L^2-derivatives coincide. Analogous L^2-differentiability results hold for $R_t(a)$ (and in the regular case $R_t(a)$ is always L^2-differentiable). As mentioned previously, we also obtain "local time at a" as the L^2-limit of certain upcrossing and downcrossing processes.

1. Statement of Results.

Recall from §0 the definition of $N_t(a)$. For $\delta > 0$ set $N_t^+(a,\delta) = N_t(a+\delta) - N_t(a)$ and $N_t^-(a,\delta) = N_t(a) - N_t(a-\delta)$. If 0 is irregular we define the "local times"

$$L_t^{(1)}(a) = |b|^{-1} \sum_{0<s\leq t} 1_{\{a\}}(X_s), \tag{1.1}$$

$$L_t^{(2)}(a) = |b|^{-1} \sum_{0\leq s<t} 1_{\{a\}}(X_s). \tag{1.2}$$

As noted in §0, $P_x(X_t = a) = 0$ for all $x, a \in \mathbf{R}$ and $t > 0$. Thus for fixed $t > 0$, $L_t^{(2)}(a) = L_t^{(1)}(a) + |b|^{-1} 1_{\{a\}}(X_0)$ almost surely.

Theorem 1. *Suppose that 0 is irregular. If $b < 0$, then for each initial distribution μ,*

$$E_\mu(\delta^{-1} N_t^+(a,\delta) - L_t^{(1)}(a))^2 \to 0, \quad \delta \downarrow 0, \tag{1.3}$$

$$E_\mu(\delta^{-1} N_t^-(a,\delta) - L_t^{(2)}(a))^2 \to 0, \quad \delta \downarrow 0. \tag{1.4}$$

If $b > 0$, then (1.3) and (1.4) hold with $L_t^{(1)}(a)$ and $L_t^{(2)}(a)$ exchanged. Also, for $i = 1, 2$,

$$\int_0^a L_t^{(i)}(x)dx = N_t(a), \quad \forall t > 0, \ \forall a \in \mathbf{R}, \ P_\mu - \text{a.s.} \tag{1.5}$$

Let $R_t^+(a,\delta)$ (resp. $R_t^-(a,\delta)$) denote the Lebesgue measure of $\{X_s : 0 \le s \le t\}$ intersected with $(a, a+\delta]$ (resp. $(a-\delta, a]$). Let $T_a = \inf\{t > 0 : X_t = a\}$, $D_a = \inf\{t \ge 0 : X_t = a\}$.

Theorem 2. *Suppose that 0 is regular. Then for each initial distribution μ,*

$$E_\mu(\delta^{-1} R_t^\pm(a,\delta) - 1_{\{T_a \le t\}})^2 \to 0, \quad \delta \downarrow 0. \tag{1.6}$$

Theorem 3. *Suppose that 0 is irregular. If $b < 0$, then for each initial distribution μ,*

$$E_\mu(\delta^{-1} R_t^+(a,\delta) - 1_{\{T_a \le t\}})^2 \to 0, \quad \delta \downarrow 0, \tag{1.7}$$

$$E_\mu(\delta^{-1} R_t^-(a,\delta) - 1_{\{D_a \le t\}})^2 \to 0, \quad \delta \downarrow 0. \tag{1.8}$$

If $b > 0$, then (1.7) and (1.8) hold with T_a and D_a exchanged.

Remark. If $\mu(\{a\}) = P_\mu(X_0 = a) = 0$, then $D_a = T_a$ and $L_t^{(1)}(a) = L_t^{(2)}(a)$, P_μ-a.s. In this case both $x \mapsto N_t(x)$ and $x \mapsto R_t(x)$ are L^2-differentiable at a.

Define $N(t) = |b|L_t^{(1)}(0)$, $N^*(t) = N(t) - 1_{\{T_0 \le t\}}$. For $x < y$ let $D_{y,x}(t)$ (resp. $U_{x,y}(t)$) be the process that counts the number of downcrossings from y to x (resp. upcrossings from x to y) completed by time t. Getoor [6] has shown that if 0 is regular, then $D_{c,a}(t)$ when properly normalized converges in L^2 to local time at 0 as $a \uparrow 0$, $c \downarrow 0$. Our last result is the irregular version of this result, which requires no normalization.

Theorem 4. *Suppose that 0 is irregular, and that $b > 0$. Then for each $t > 0$, each of the processes $D_{c,0}(t)$, $D_{c,a}(t)$, $U_{a,0}(t)$, $U_{a,c}(t)$ converges in $L^2(P_0)$ to $N(t)$ as $a \uparrow 0$, $c \downarrow 0$. If $x \neq 0$, then $D_{c,0}(t)$ and $D_{c,a}(t)$ converge in $L^2(P_x)$ to $N(t)$, while $U_{a,0}(t)$ and $U_{a,c}(t)$ converge in $L^2(P_x)$ to $N^*(t)$, as $a \uparrow 0$, $c \downarrow 0$.*

Of course, the analogous result holds if $b > 0$—simply apply Theorem 4 to $-X$. In Theorem 4, the anomalous behavior of the upcrossing processes when $X_0 = x \neq 0$ is explained by the following story, which is true with probability approaching 1 as $a \uparrow 0$, $c \downarrow 0$. (See Lemma 7 in §2.) Let a and c be close to 0, with $a < 0 < c$. Starting at x, our process X hits c, 0, a (in that order), thereby initiating the first upcrossing. The process X then returns to c (before hitting 0 or a) completing the first a to c upcrossing; X then hits 0, followed by a. In short, *two* hits of 0 are "used up" by the first a to c upcrossing.

We note that if X is transient, then Theorems 1–4 remain valid for $t = +\infty$, and only slight alterations of our proofs are required.

2. Proofs.

Our blanket hypotheses ($\{0\}$ is non-polar, X is neither a compound Poisson process, nor a compound Poisson process with drift) ensure the following (see [4]). First, Lebesgue measure is a reference measure for X. Second, the λ-potential kernel density $g^\lambda(x,y) = g^\lambda(y-x)$ satisfies

(i) g^λ is bounded, and continuous on $\mathbf{R}\backslash\{0\}$;

(ii) g^λ is continuous at 0 if and only if 0 is regular (for $\{0\}$);

(iii) if 0 is irregular, then $g^\lambda(0+)$ and $g^\lambda(0-)$ exist, and $g^\lambda(0+) = g^\lambda(0) < g^\lambda(0-)$ if $b < 0$, while $g^\lambda(0-) = g^\lambda(0) < g^\lambda(0+)$ if $b > 0$;

(iv) defining

$$h^\lambda(x) = E_0(e^{-\lambda T_x}),$$

we have $h^\lambda = c^\lambda g^\lambda$, where $c^\lambda = [g^\lambda(0+) \vee g^\lambda(0-)]^{-1}$;

(v) if 0 is irregular, $g^\lambda(0+) - g^\lambda(0-) = b^{-1}$ (see the appendix of [10]).

Combining (iv) and (v) we obtain

$$[1 - h^\lambda(0)]/c^\lambda = 1/|b|. \tag{2.1}$$

The above statements hold if $\lambda > 0$, and also if $\lambda = 0$ when X is transient.

For the proof of Theorem 1 we introduce the 1-subprocess $Y = (Y_t : t \geq 0)$, obtained by killing X at an independent exponential time S of mean 1. The potential kernel of Y is $g^1(x,y) = g^1(y-x)$, and the probability that Y hits a, given $Y_0 = x$, is $h^1(a-x)$. We can (and do) assume that the probability space on which X and Y are defined supports shift operators $\bar{\theta}_t$ for Y: $Y_{t+s} = Y_s \circ \bar{\theta}_t$. (As usual, the shifts for X are denoted θ_t.)

When 0 is irregular we defined in §1 two versions of "local time at a", namely $L_t^{(1)}(a)$ and $L_t^{(2)}(a)$. We now define a hybrid local time $L_t(a)$ by setting $L_t(a) = L_t^{(1)}(a)$ if $b < 0$, $L_t(a) = L_t^{(2)}(a)$ if $b > 0$. Local time at a for the process Y is thus $\ell_t(a) = L_{t\wedge S}(a)$. For a Borel set $B \subset \mathbf{R}$ we define the occupation time processes

$$N_t(B) = \int_0^t 1_B(X_s)ds,$$

$$M_t(B) = N_{t\wedge S}(B) = \int_0^t 1_B(Y_s)ds,$$

and set $\ell(a) = \ell_\infty(a)$, $M(B) = M_\infty(B)$, etc. Clearly $N_t(B)$ (resp. $M_t(B)$) is a continuous additive functional of X (resp. Y) with λ-potential $\int_B g^\lambda(y - \cdot)dy$ (resp. $\int_B g^{1+\lambda}(y - \cdot)dy$).

Lemma 1. *If 0 is irregular, then*

$$E_x(\ell(a)) = g^1((a-x)+),$$

and

$$E_x(\ell(a)^2) = g^1((a-x)+)[g^1(0+)+g^1(0-)].$$

Proof. Observe that under P_x, the law of $\ell(a)$ is that of $W(Z+1)/|b|$, where W and Z are independent, W is an indicator random variable with success probability $h^1((a-x)+)$, and Z has the geometric distribution with parameter $h^1(0)$ (i.e., $P_x(Z=n) = [1-h^1(0)][h^1(0)]^n$, $n \geq 0$). The lemma now follows easily. □

For the next lemma let $M^+(a,\delta) = M((a,a+\delta]) = N_S^+(a,\delta)$, $\delta > 0$.

Lemma 2. *If* 0 *is irregular, then*

(2.2)

$$E_x(M^+(a,\delta)\ell(a)) = \int_a^{a+\delta} [g^1((a-x)+)g^1(y-a) + g^1((a-y)+)g^1(y-x)]dy,$$

and

$$E_x(M^+(a,\delta)^2) = 2\int_a^{a+\delta}\int_a^{a+\delta} g^1(z-x)g^1(y-z)dydz. \tag{2.3}$$

Proof. Suppose that $b < 0$. Then $t \mapsto \ell_t(a) = L^{(1)}_{t\wedge S}(a)$ is a right continuous additive functional of Y, with a jump $1/|b|$ at each $s > 0$ for which $Y_s = a$. Writing $B = (a, a+\delta]$ we therefore have

$$E_x(\ell(a)M^+(a,\delta)) = I_1 + I_2,$$

where

$$I_1 = E_x\left(\iint\limits_{0<s\leq t} d\ell_s(a)1_B(Y_t)dt\right),$$

and

$$I_2 = E_x\left(\iint\limits_{0<t<s} d\ell_s(a)1_B(Y_s)dt\right).$$

Using Lemma 1, we obtain

$$\begin{aligned} I_1 &= E_x\left(\int_{(0,\infty)} d\ell_s(a) E_{Y_s}(M(b))\right) \\ &= E_x\left(\int_{(0,\infty)} d\ell_s(a)\right) E_a(M(B)) \\ &= g^1((a-x)+) \int_B g^1(y-a)dy, \end{aligned}$$

and similarly

$$I_2 = \int_a^{a+\delta} g^1((a-y)+)g^1(y-x)dy.$$

Thus (2.2) follows for $b < 0$, and the case $b > 0$ is handled likewise. As to (2.3),

$$\begin{aligned} E_x(M(B)^2) &= E_x\left(\int_0^\infty \int_0^\infty 1_B(Y_s)1_B(Y_t)ds\, dt\right) \\ &= 2\, E_x\left(\int_0^\infty 1_B(Y_s)ds \int_s^\infty 1_B(Y_t)dt\right) \\ &= 2\, E_x\left(\int_0^\infty 1_B(Y_s)E_{Y_s}(M(B))ds\right) \\ &= 2\int_B g^1(y-x)\left[\int_B g^1(z-y)dz\right] dy, \end{aligned}$$

which yields (2.3) if $B = (a, a+\delta]$. □

Lemma 3. *If* 0 *is irregular then*

$$\lim_{\delta\downarrow 0} E_x(\delta^{-1}M^+(a,\delta)\cdot \ell(a)) = \lim_{\delta\downarrow 0} E_x(\delta^{-2}M^+(a,\delta)^2) = E_x(\ell(a)^2).$$

Proof. This follows immediately from Lemmas 1 and 2, and the properties of g^1 listed earlier. □

Proof of Theorem 1. From Lemma 3 we deduce that for each x,

$$E_x(\delta^{-1}M^+(a,\delta) - \ell(a))^2 \to 0, \quad \delta \downarrow 0. \tag{2.4}$$

Since g^1 is bounded, the expectation in (2.4) is bounded uniformly for $0 < \delta \leq 1$, $x \in \mathbf{R}$. Thus, by the bounded convergence theorem, for any initial distribution μ,

$$E_\mu(\delta^{-1}M^+(a,\delta) - \ell(a))^2 \to 0, \quad \delta \downarrow 0. \tag{2.5}$$

In particular, if ν denotes the P_μ-law of Y_t, then

$$\begin{aligned} &E_\mu([\delta^{-1}M^+(a,\delta)-\ell(a)]^2\circ\bar{\theta}_t) \\ &\qquad = E_\nu(\delta^{-1}M^+(a,\delta)-\ell(a))^2 \to 0, \quad \delta\downarrow 0. \end{aligned} \tag{2.6}$$

But

$$M_t^+(a,\delta) = M^+(a,\delta) - M^+(a,\delta)\circ\bar{\theta}_t,$$

and

$$\ell_t(a) = \ell(a) - \ell(a)\circ\bar{\theta}_t,$$

where $M_t^+ = N_{t\wedge S}^+$. Combining (2.5) and (2.6) we obtain

$$\delta^{-1}M_t^+(a,\delta) \to \ell_t(a) \quad \text{in} \quad L^2(P_\mu), \quad \delta\downarrow 0.$$

Finally, since $M_t^+(a,\delta) = N_{t\wedge S}^+(a,\delta)$, $\ell_t(a) = L_{t\wedge S}(a)$,

$$\begin{aligned} 0 &\le P_\mu(S>t)E_\mu(\delta^{-1}N_t^+(a,\delta)-L_t(a))^2 \\ &\qquad \le E_\mu(\delta^{-1}M_t^+(a,\delta)-\ell_t(a))^2 \to 0, \ \delta\downarrow 0, \end{aligned}$$

which forces (1.3). The relation (1.4) obtains upon replacing X by $-X$. Finally, for fixed $a>0$, both sides of (1.5) are continuous additive functionals of X with the same finite 1-potential $\int_0^a g^1(y-\cdot)dy$ (cf. Lemma 1). By the uniqueness theorem for continuous additive functionals, (1.5) holds $\forall t>0$, P_μ-a.s., for any fixed $a>0$. The case $a\le 0$ is handled in exactly the same way. Since both sides of (1.5) are evidently continuous in a, the exceptional set can be chosen independently of a as well. □

Proof of Theorem 2. Since 0 is regular, $x\mapsto h^\lambda(x) = P_0(e^{-\lambda T_x})$ is continuous, for each $\lambda>0$. Moreover, since X_t has a continuous distribution, $P_x(T_a=t)\le P_x(X_t=a)=0$, $t>0$. Thus, by the continuity theorem for Laplace transforms, $(x,y)\mapsto P_x(T_y\le t) = P_0(T_{y-x}\le t)$ is continuous. In particular, by bounded convergence, $y\mapsto P_\mu(T_y\le t)$ is continuous for each initial law μ. Let $T_x^* = T_x \circ \theta_{T_a}$. Now $0 \le \delta^{-1} R_t^+(a,\delta) \le 1$, and so

$$\begin{aligned}
P_\mu(T_a \le t) &\ge \delta^{-1} P_\mu(R_t^+(a,\delta);\ T_a \le t) \\
&= \delta^{-1} \int_a^{a+\delta} P_\mu(T_x \le t, T_a \le t)dx \\
&\ge \delta^{-1} \int_a^{a+\delta} P_\mu(T_x^* \le t, T_a \le t)dx \\
&= \delta^{-1} \int_a^{a+\delta} \left[\int_0^t P_\mu(T_a \in ds) P_a(T_x \le t-s)\right] dx \\
&\underset{\delta \downarrow 0}{\longrightarrow} \int_0^t P_\mu(T_a \in s) P_a(T_a \le t-s) \\
&= P_\mu(T_a \le t),
\end{aligned}$$

since $P_a(T_a = 0) = 1$, a being regular for $\{a\}$. Thus

$$E_\mu(\delta^{-1} R_t^+(a,\delta); T_a \le t) \to P_\mu(T_a \le t), \quad \delta \downarrow 0, \tag{2.7}$$

and in the same way

$$E_\mu(\delta^{-1} R_t^+(a,\delta)) \to P_\mu(T_a \le t), \quad \delta \downarrow 0. \tag{2.8}$$

Combining (2.7) and (2.8) with $0 \le \delta^{-1} R_t^+(a,\delta) \le 1$, we obtain the "+" version of (1.6). The "−" version follows by substituting $-X_t$ for X_t. □

We shall prove Theorems 3 and 4 only in the case $b < 0$. The case $b > 0$ follows, as before, upon replacing X by $-X$. *Thus for the rest of the paper we assume that* 0 *is irregular for* $\{0\}$ *and that* $b < 0$.

The proof of Theorem 3 requires a few more lemmas.

Lemma 4. *For all* $x, a \in \mathbf{R}$,

$$\lim_{c \downarrow a} P_x(T_c < T_a \le t) = P_x(T_a \le t); \tag{2.9}$$

$$\lim_{c \downarrow a} P_x(T_a < T_c \le t) = 0; \tag{2.10}$$

$$\lim_{c \downarrow a} P_x(T_c \le t) = P_x(T_a \le t). \tag{2.11}$$

Proof. Spatial invariance allows us to consider only the case $x = 0$. Let $H^\lambda(a,c) = P_0(e^{-\lambda T_c}; T_a < T_c)$, $\gamma^\lambda(a,c) = P_0(e^{-\lambda T_a}; T_a < T_c)$. Then

$$H^\lambda(a,c) = \gamma^\lambda(a,c) h^\lambda(c-a),$$

and for $a \neq c$,

$$h^\lambda(a) = \gamma^\lambda(a,c) + \gamma^\lambda(c,a)h^\lambda(a-c),$$

$$h^\lambda(c) = \gamma^\lambda(a,c)h^\lambda(c-a) + \gamma^\lambda(c,a).$$

Solving these last two equations we obtain

$$\gamma^\lambda(c,a) = [h^\lambda(c) - h^\lambda(a)h^\lambda(c-a)][1 - h^\lambda(a-c)h^\lambda(c-a)]^{-1}. \tag{2.12}$$

Since $b < 0$, h^λ is right continuous, and so

$$\lim_{c \downarrow a} h^\lambda(c) = h^\lambda(a) = P_0(e^{-\lambda T_a}),$$

which implies (2.11) by the continuity theorem for Laplace transforms. Similarly, (2.9) follows by using (2.12):

$$\begin{aligned}\lim_{c\downarrow a} h^\lambda(c,a) &= \lim_{c\downarrow a} \gamma^\lambda(c,a)h^\lambda(a-c) \\ &= [h^\lambda(a) - h^\lambda(a)h^\lambda(0)][1 - h^\lambda(0)]^{-1}h^\lambda(0-) \\ &= h^\lambda(a) = P_0(e^{-\lambda T_a}).\end{aligned}$$

Finally, (2.10) follows upon subtracting (2.9) from (2.11), since $\{T_a < T_c \le t\} \subset \{T_c \le t\}\backslash\{T_c < T_a \le t\}$. □

The proof of Lemma 5 below is essentially the same as that of Lemma 4. We omit the details.

Lemma 5. *For all* $x, a \in \mathbf{R}$,

$$\lim_{c\uparrow a} P_x(T_c < T_a \le t) = 1_{\{a=x\}};$$

$$\lim_{c\uparrow a} P_x(T_a < T_c \le t) = 1_{\{a\neq x\}}P_x(T_a \le t);$$

$$\lim_{c\uparrow a} P_x(T_c \le t) = P_x(D_a \le t).$$

Proof of Theorem 3. Since $0 \le R_t^+(a,\delta) \le \delta$, we have

$$0 \le E_x(\delta^{-1}R_t^+(a,\delta) - 1_{\{T_a \le t\}})^2 \tag{2.13}$$

$$\le \delta^{-1}\int_a^{a+\delta} P_x(T_y \le t))dy - 2\delta^{-1}\int_a^{a+\delta} P_x(T_y \le t, T_a \le t)dy + P_x(T_a \le t).$$

But $P_x(T_y \le t, T_a \le t) = P_x(T_y < T_a \le t) + P_x(T_a < T_y \le t) \to P_x(T_a \le t)$ as $y \downarrow a$ (Lemma 4). A second application of Lemma 4 shows that the extreme right-hand term in (2.13) tends to 0 as $\delta \downarrow 0$. Thus (1.7) holds if μ is the point mass at x; (1.7) for arbitrary μ follows by bounded convergence. Exactly the same argument (Lemma 5 replacing Lemma 4) yields (1.8). □

For the proof of Theorem 4 we need some facts concerning delayed renewal processes. Let W_1 $W_2, \ldots$ be independent $(0,\infty]$-valued random variables such that W_1 has law F_1, and such that W_2, $W_3, \ldots$ have law F_2. The associated renewal counting process is $N(t) = \sum_{n\ge 1} 1_{\{S_n \le t\}}$, where $S_n = W_1 + \cdots + W_n$. If F_1 and F_2 are continuous distributions (except perhaps for a mass at ∞), then $t \to N(t)$ is stochastically continuous, and for any $p \ge 1$, $t \to E(N(t)^p)$ is finite and continuous (see (2.14) below). If F is a distribution function, then $\hat{F}$ denotes its Laplace transform.

Lemma 6. *Let $N_1(t)$, $N_2(t), \ldots$ be delayed renewal processes with interarrival laws F_{n1}, F_{n2} $(n \ge 1)$. Let $N(t)$ be a delayed renewal process with interarrival laws F_1, F_2. Suppose that F_1 and F_2 are continuous on $(0,\infty)$ and that $\hat{F}_{ni}(\lambda) \to \hat{F}_i(\lambda)$, $n \to \infty$, for $i = 1,2$, $\lambda > 0$. Then*

(i) $N_n(t) \xrightarrow{D} N(t)$, $n \to \infty$ $(t > 0)$,

(ii) $E(N_n(t)^p) \to E(N(t)^p)$, $n \to \infty$, $(t > 0,\ p = 1, 2, 3 \ldots)$.

Proof. Let the renewal epochs of $N_n(t)$ be $\{S_{ni}: i \ge 1\}$. Our hypotheses clearly imply that $S_{ni} \xrightarrow{D} S_i$, $n \to \infty$ $(i \ge 1)$. Thus $N_n(t) \xrightarrow{D} N(t)$, since $\{N_n(t) \ge k\} = \{S_{nk} \le t\}$, $\{N(t) \ge k\} = \{S_k \le t\}$ and $t \mapsto P(S_k \le t)$ is continuous on $(0,\infty)$. If $M(t)$ is an arbitrary delayed renewal process with interarrival laws G_1, G_2, then

$$\int_0^\infty e^{-\lambda t} dt\, E(e^{-uM(t)}) \tag{2.14}$$

$$= \lambda^{-1}(1 - \hat{G}_1(\lambda)) + \lambda^{-1}\hat{G}_1(\lambda)(1 - \hat{G}_2(\lambda))e^{-u}[1 - e^{-u}\hat{G}_2(\lambda)]^{-1}.$$

Differentiation of (2.14) at $u = 0$ (p times) reveals that $E(M(t)^p) < \infty$ for all $t > 0$, and that $\int_0^\infty e^{-\lambda t} E(M(t)^p) dt$ is finite and continuous as a function of

$\hat{G}_1(\lambda), \hat{G}_2(\lambda)$. Assertion (ii) now follows from the continuity theorem for Laplace transforms, since $t \mapsto E(N(t)^p)$ is increasing and continuous. □

Lemma 7. *For $t > 0$, $x \neq 0$,*

(i) $\lim_{z\to 0} P_0(T_0 < T_z, T_0 \leq t) = 0$;

(ii) $\lim_{a\uparrow 0, c\downarrow 0} P_c(T_a < T_0, T_a \leq t) = 0$;

(iii) $\lim_{a\uparrow 0} P_x(T_a < T_0, T_a \leq t) = 0$.

Proof. Rewriting (2.12),

$$E_x(e^{-T_y};\, T_y < T_z) = \frac{h^\lambda(y-x) - h^\lambda(z-x)h^\lambda(y-z)}{1 - h^\lambda(y-z)h^\lambda(z-y)}. \tag{2.15}$$

Since $b < 0$, $1 = h^\lambda(0-) > h^\lambda(0+) = h^\lambda(0)$, and so taking $x = y = 0$ in (2.15) we see that

$$\begin{aligned}\lim_{z\to 0} E_0(e^{-\lambda T_0};\, T_0 < T_z) &= [h^\lambda(0) - h^\lambda(0+)h^\lambda(0-)][1 - h^\lambda(0+)h^\lambda(0-)]^{-1}\\ &= 0,\end{aligned}$$

which implies (i). Points (ii) and (iii) follow similarly. □

Henceforth $N(t)$ denotes the delayed renewal counting process that counts the returns of X to 0. Recall that $N^*(t) = N(t) - 1_{\{T_0 \leq t\}}$. If S and T are stopping times, then $T \circ S$ denotes the stopping time $S + T \circ \theta_S$.

Lemma 8. *For $t > 0$, $x \in \mathbf{R}$,*

$$D_{c,0}(t) \xrightarrow{P_x} N(t), \quad c \downarrow 0.$$

Proof. Let $S_0 = 0$, and let $0 < S_1 < S_2 \cdots$ be the times when X returns to 0. Obviously $D_{c,0}(t) \leq N(t)$, while for $n \geq 1$,

$$\begin{aligned}\{N(t) > D_{c,0}(t), N(t) = n\} &\subset \bigcup_{i=0}^{n-1} \{T_c \circ S_i > S_{i+1}, S_n \leq t\}\\ &\subset \bigcup_{i=0}^{n-1} \{T_c \circ S_i > T_0 \circ S_i, T_0 \circ S_i \leq t\}.\end{aligned}$$

It follows that

$$P_x(N(t) > D_{c,0}(t), N(t) = n) \leq nP_0(T_c > T_0, T_0 \leq t), \tag{2.16}$$

and the R.H.S. of (2.16) tends to 0 as $c \downarrow 0$ by Lemma 7(i). But

$$\begin{aligned} P_x(N(t) > D_{c,0}(t)) \leq P_x(N(t) > D_{c,0}(t), N(t) \leq n) \\ + P_x(N(t) > n), \end{aligned} \tag{2.17}$$

so letting $c \downarrow 0$ and then $n \to \infty$ in (2.17) we conclude that $P_x(N(t) > D_{c,0}(t)) \to 0$ as $c \downarrow 0$, and the lemma is proved. □

Lemma 9. *For $x \neq 0$,*

$$U_{a,0}(t) \xrightarrow{P_x} N^*(t), \quad a \uparrow 0,$$

while for $x = 0$,

$$U_{a,0}(t) \xrightarrow{P_0} N(t), \quad a \uparrow 0.$$

Proof. We omit the proof of the case $x = 0$, which is the same as that of Lemma 8. For $x \neq 0$, we have (using the notation of Lemma 8),

$$\begin{aligned} &P_x(N^*(t) > U_{a,0}(t), N^*(t) = n) \\ &\quad \leq P_x(T_0 \circ T_0 < T_a \circ T_0, T_0 \circ T_0 \leq t) \\ &\qquad + \sum_{i=1}^{n-1} P_x(S_{i+1} < T_a \circ S_i, S_{i+1} \leq t) \\ &\quad \leq P_0(T_0 < T_a, T_0 \leq t) \\ &\qquad + (n-1)P_0(T_0 < T_a, T_0 \leq t), \end{aligned}$$

and the last term above tends to 0 as $a \uparrow 0$ (Lemma 7(i)). As in the proof of Lemma 8, we conclude that

$$P_x(N^*(t) > U_{a,0}(t)) \to 0, \quad a \uparrow 0.$$

On the other hand,

$$\begin{aligned} P_x(U_{a,0}(t) > N^*(t)) &\leq P_x(U_{a,0}(t) \geq N(t)) \\ &\leq P_x(T_a < T_0, T_a \leq t) \to 0, \end{aligned}$$

as $a \uparrow 0$, by Lemma 7(iii). □

Lemma 10. *As $a \uparrow 0$, $c \downarrow 0$,*

$$D_{c,a}(t) - D_{c,0}(t) \to 0, \quad U_{a,c}(t) - U_{a,0}(t) \to 0$$

in P_x-probability, for each $x \in \mathbf{R}$.

Proof. We consider only the downcrossing case. Let $\sigma_0 = 0$, and let $0 < \sigma_1 < \sigma_2 < \cdots$ denote the times at which the c to a downcrossings are completed. For $n \geq 1$ we have

$$\begin{aligned} &P_x(D_{c,a}(t) > D_{c,0}(t), D_{c,a}(t) = n) \\ &\leq \sum_{i=0}^{n-1} P_x(\sigma_{i+1} < T_0 \circ T_c \circ \sigma_i, \sigma_{i+1} \leq t) \\ &\leq n\, P_c(T_a < T_0, T_a \leq t) \to 0, \quad a \uparrow 0, c \downarrow 0, \end{aligned} \tag{2.18}$$

by Lemma 7(ii). Now $D_{c,a}$ is a delayed renewal process with interarrival laws $F_{1,c,a}$ and $F_{2,c,a}$, where $F_{1,c,a}$ (resp. $F_{2,c,a}$) is the law of $T_a \circ T_c$ under P_x (resp. P_a). Thus $\hat{F}_{1,c,a}(\lambda) = h^\lambda(c-x)h^\lambda(a-c)$ and $\hat{F}_{2,c,a}(\lambda) = h^\lambda(c-a)h^\lambda(a-c)$, so that $\hat{F}_{1,c,a}(\lambda) \to h^\lambda(-x) = P_x(e^{-\lambda T_0})$, $\hat{F}_{2,c,a}(\lambda) \to h^\lambda(0) = P_0(e^{-\lambda T_0})$ as $a \uparrow 0$, $c \downarrow 0$. Consulting Lemma 6 we see that

$$E_x(D_{c,a}(t)) \to E_x(N(t)) = |b|E_x(L_t^{(1)}(0)) < \infty,$$

as $a \uparrow 0$, $c \downarrow 0$. Hence there are constants $K > 0$, $\delta > 0$ (depending on x and t perhaps) such that

$$\sup_{\substack{0<c<\delta \\ -\delta<a<0}} E_x(D_{c,a}(t)) \leq K,$$

and so

$$\sup_{\substack{0<c<\delta \\ -\delta<a<0}} P_x(D_{c,a}(t) > n) \leq K/n \to 0, \quad \text{as} \quad n \to \infty.$$

This last result, in conjunction with (2.18), allows us to conclude that

$$P_x(D_{c,a}(t) > D_{c,0}(t)) \to 0, \quad a \uparrow 0, c \downarrow 0.$$

In much the same way $P_x(D_{c,a}(t) < D_{c,0}(t)) \to 0$ as $a \uparrow 0$, $c \downarrow 0$. □

Proof of Theorem 4. Since convergence in probability coupled with convergence of second moments implies L^2 convergence, Theorem 4 follows at once from Lemmas 6–10. □

REFERENCES

1. V. BALLY. Approximation theorems for the local time of a Markov process. *Studii si Cercetări Matematice* **38** (1986) 139–147. Bucuresti.
2. V. BALLY and L. STOICA. A class of Markov processes which admit local times. *Ann. Probab.* **15** (1987) 241–262.
3. R. M. BLUMENTHAL and R. K. GETOOR. Local times for Markov processes. *Z. Wahrsch. verw. Gebiete* **3** (1964) 50–74.
4. J. BRETAGNOLLE. Resultats de Kesten sur les processus a acroissements indépendants. *Sem. Prob.* V. *Lecture Notes in Math.* **191** 21–36, Springer, Berlin, 1971.
5. B. E. FRISTEDT and S. J. TAYLOR. Constructions of local time for a Markov process. *Z. Wahrsch. verw. Gebiete* **62** (1983) 73–112.
6. R. K. GETOOR. Another limit theorem for local time. *Z. Wahrsch. verw. Gebiete* **34** (1976) 1–10.
7. R. J. GRIEGO. Local time as a derivative of occupation times. *Ill. J. Math.* **11** (1967) 53–64.
8. G. A. HUNT. Markoff processes and Potentials III. *Ill. J. Math.* **2** (1958) 151–213.
9. S.C. PORT and C. J. STONE. Infinitely divisible processes and their potential theory I. *Ann. Inst. Fourier* **21** (1971) 157–275.
10. S. C. PORT. Stable processes with drift on the line. *Trans. Amer. Math. Soc.* (to appear)

P. J. FITZSIMMONS
Department of Mathematics, C-012
University of California, San Diego
La Jolla, CA 92093

S. C. PORT
Department of Mathematics
University of California
Los Angeles, CA 90024

MARTINGALE PROBLEMS ASSOCIATED WITH THE BOLTZMANN EQUATION

by

J. HOROWITZ*) and R. L. KARANDIKAR

1. INTRODUCTION

The Boltzmann equation is a nonlinear integro-partial differential equation that is supposed to describe the distribution of positions and velocities of the molecules in a dilute gas as a function of time. It is assumed that only two molecules at a time can collide.

If the precollision velocities of two molecules are u, v in $\mathbb{R}^3$ and the postcollision velocities are u', v', we write $(u,v) \to (u',v')$ to denote the collision. Because of conservation of momentum and energy, u' and v' lie on the sphere S_{uv} with north pole at u and south pole at v. Thus u', v' are specified by colatitude $0 < \theta \leq \pi$ and longitude $0 \leq \phi < 2\pi$. We write $A = (0,\pi] \times [0,2\pi)$. There is necessarily some arbitrariness in the choice of ϕ that causes complications later.

The "classical" Boltzmann equation (abbreviation: B.E.) assumes that the position-velocity distribution is given by a density $f(t,x,v)$. In the *spatially homogeneous* case that we consider, there is no dependence on x. For a discussion of the physical meaning of the B.E. as well as related terminology, see Truesdell and Muncaster [24],

*) Research partially supported by NSF Grant DMS-8602651 and an Indo-American Fellowship under the Indo-U.S. Subcommission.

Cercignani [2], and Thompson [23]. Our view of the B.E. will be a dispassionate, mathematical one, like that of [24], p. 93.

Instead of the classical equation we consider the *weak, spatially homogeneous Boltzmann equation*

$$(1.1) \qquad \frac{d}{dt}\langle \nu_t, f\rangle = \langle \nu_t \otimes \nu_t, Kf\rangle, \quad f \in C_c^2(\mathbb{R}^3),$$

where $\langle \nu, f\rangle$ means $\int f d\nu$, $\otimes$ denotes a product of measures, K is the operator

$$(1.2) \qquad Kf(u,v) = \frac{1}{2\pi}\int_A (f(u') - f(u))d\phi Q(d\theta), \quad f \in C_c^2(\mathbb{R}^3)$$

(involving the collision $(u,v) \to (u',v')$), and the "unknown" is the family $\{\nu_t\colon t \geq 0\}$ of probability measures on $\mathbb{R}^3$. In (1.2), Q is a measure on $(0,\pi]$ that governs the physics of the interactions between molecules, always assumed to satisfy

$$(1.3) \qquad \int_0^\pi \theta^2 Q(d\theta) < \infty ,$$

or, equivalently, (2.8) below. That (1.1) is an appropriate weak form of the B.E. (*weak, spatially homogeneous* are understood henceforth) is shown by Tanaka [20].

We regard (1.1) from the point of view of martingale problems, along lines related to Tanaka [21], Funaki [7], Sznitman [19], and Oelschläger [16].

The main contributions of this paper, which are discussed in more detail in the rest of this section, are

a) existence and uniqueness (in law) of the n-particle process approximating the infinite particle gas;

b) convergence of the n-particle Boltzmann process (so-called) to the infinite particle Boltzmann process;

c) existence and uniqueness for the B.E. assuming only a finite second moment, for Maxwellian-type molecules;

d) elucidation of the time-inhomogeneous Markov processes of McKean [15] and Tanaka [20];

e) trend to equilibrium and other properties of solutions of the B.E.

DESCRIPTION OF RESULTS

The B.E. is an infinite particle model for a dilute gas consisting of a large but finite number of particles. Our approach is to approximate the infinite particle system by a system of n particles and let $n \to \infty$. The fact that this is possible lends support to the use of the B.E. as a model for real physical systems.

Let $V^n(t) = (V_1^n(t),\dots,V_n^n(t))$ be the vector of velocities at time t for n particles. We shall regard V^n as a stochastic process, the randomness in which can be considered to arise either because of our (self-imposed) ignorance of the particles' positions or simply as an artifice: according to classical physics the motion is deterministic.

In §2 we prove existence and uniqueness in law for V^n. It is a Markov process with generator given formally by

$$(1.4) \qquad K_n f(\underline{u}) = \frac{1}{2\pi n} \sum_{1\le i<j\le n} \int_A (f(u_1,\dots,u_i',\dots,u_j',\dots,u_n) - f(\underline{u}))d\phi Q(d\theta),$$

$f \in C_c^2((\mathbf{R}^3)^n)$, $\underline{u} = (u_1,\dots,u_n) \in (\mathbf{R}^3)^n$, and where the (i,j)-term involves the collision $(u_i,u_j) \to (u_i',u_j')$ with angular coordinates (θ,ϕ).

When $Q(0,\pi] < \infty$, as in Kac [11], Grünbaum [9], the existence is clear, but not when $Q(0,\pi] = \infty$, as V^n must then be a pure jump Markov process with all instantaneous states. Because use of the Hille-Yosida theorem would require precise determination of the domain of K_n, we turn instead to the martingale problem formulation - Kac's "master equation" in modern dress - where the existence is readily obtained.

NOTE. Since there will be many references to the books of Jacod [10] and Ethier and Kurtz [4], we cite them henceforth as [J] and [EK]. Any unexplained terminology can be found there. Also, we use the following notation: $\mathcal{B}(E)$ for the Borel σ-field on a topological space E, $\mathcal{P}(E)$ the space of probability measures (weak topology) on $\mathcal{B}(E)$, $\xrightarrow{w}$ weak convergence, $\mathcal{L}(\cdot)$ for the probability law of a rv or stochastic process, $\xrightarrow{D}$ convergence in law, and, finally, := for "equal by definition".

For uniqueness, the results of Stroock [17], Lepeltier-Marchal [13], Sznitman [19] (appendix), Komatsu [12], and Bass [1] appear to be insufficient for our situation, so we use the equivalence between martingale problems and stochastic differential equations (SDEs) in [J], chs. XIII, XIV. The SDE here is of pure jump type with non-Lipschitz coefficients. The uniqueness proof is then based on an idea of Tanaka [20], approximating V^n by simpler processes for which uniqueness in law can be shown. As a corollary, V^n is a time-homogeneous Markov process.

Let ν_t^n be the empirical measure of $V^n(t)$,

$$\nu_t^n = \frac{1}{n} \sum_{i=1}^{n} \delta_{V_i^n(t)},$$

and put $U_t^n = V_1^n(t)$. The *n-particle Boltzmann process* is defined in §3 as $\beta_t^n := (U_t^n, \nu_t^n)$. This amounts to watching a tagged particle in a bath of $n-1$ like particles. Let $\mathcal{P}_2 = \mathcal{P}_2(\mathbb{R}^3)$ be the space of probability measures μ on $\mathbb{R}^3$ such that $\int |v|^2 \mu(dv) < \infty$, with topology determined by the ρ_2-metric of Tanaka [20], i.e., $\mu_n \xrightarrow{\rho_2} \mu$

if $\mu_n \xrightarrow{w} \mu$ and $\int|v|^2 d\mu_n \to \int|v|^2 d\mu$. Then β^n is a time-homogeneous Markov process with state space $E = \mathbb{R}^3 \times P_2$, hence it is also a solution of the L_n-martingale problem, L_n being the generator restricted to a convenient domain.

The main result of §3 is that, if $L(\beta_0^n) \xrightarrow{w} \mu \in P(E)$, then $\beta^n \xrightarrow{D} \beta$ in $D_E[0,\infty)$, $\beta_t := (U_t, \nu_t)$, and β solves the L-local martingale problem with initial law μ, where $L = \lim_n L_n$ in an appropriate sense. Such a process β is called a(n) *(infinite particle) Boltzmann process*. It is important to note that the L-local martingale problem is a "classical" time-homogeneous one as in [EK], ch. 4.

Section 4 deals with uniqueness for the L-local martingale problem and for the B.E. itself. Again we rely on SDEs, this time for the velocity component U_t of the Boltzmann process β_t, which are equivalent to the "nonlinear" martingale problems of Funaki [5], [6], [7], and also to the SDEs of Tanaka [22]. Tanaka's results then imply that $L(U_\cdot)$ is the unique solution to Funaki's martingale problem.

Using this we carry out a suggestion made in [6] to prove uniqueness for the B.E. in the class $D_{P_2}[0,\infty)$; in fact the solutions are in $C_{P_2}[0,\infty)$.

According to [6], p.348, a uniqueness proof could be based on a result of Echeverria [3]. We did not see how to complete this argument, but the necessary result is given in [EK] for the time-homogeneous case. We require the time-inhomogeneous case, and give the result in App. B. Thus we prove existence and uniqueness for the B.E. for Maxwellian (hence non-cut-off) molecules assuming that ν_0 (see eq. (1.1)) has a finite second moment. By comparison, Sznitman [19] requires a finite third moment and a cut-off interaction, but allows some velocity-dependent interactions, e.g., hard spheres.

Uniqueness for the B.E. readily implies it also for the L-local martingale problem, hence the L-local martingale problem is well-posed, and we get a number of interesting corollaries. The first is that $\beta_t = (U_t, \nu_t)$ is a *time-homogeneous* Markov process and ν_t is deterministic, except for a possibly random initial value ν_0 (i.e., ν_t is F_0^β-measurable), and solves the B.E. It follows that U_t is a time-inhomogeneous Markov process of the type considered by Tanaka [20]. This makes the nonhomogeneity of the Markov processes of McKean [15] and Tanaka completely transparent.

The convergence $\nu_\cdot^n \xrightarrow{D} \nu_\cdot$ (see §4) is much stronger than classical propagation of chaos (e.g., [9]): we do not assume symmetry and get convergence of the processes rather than the marginals. We see, therefore, that every solution of the B.E. is "physical" in that it can be approximated by n-particle distributions. Sznitman [19] obtains a similar "functional" propagation of chaos assuming symmetry and Q finite; Oelschläger [16] also has such a result, without symmetry, but under Lipschitz conditions not met by the B.E.

Suppose the laws $L(V^n(0))$ are symmetric; then $\nu_0^n \xrightarrow{D} \nu_0$ iff the random vectors $(V_1^n(0),\dots,V_n^n(0),0,0,\dots) \xrightarrow{D} (U_1,U_2,\dots)$ in $(\mathbf{R}^3)^\infty$, where $\{U_i\}$ is i.i.d. given ν_0. If ν_0 is nonrandom, then $\nu_0 = L(U_i)$: this is the classical case. We then have $\nu_t = L(U_t)$ and U_t is the nonhomogeneous Markov process of McKean type [15]. If ν_0 is random, then $\nu_t = L(U_t|\nu_0)$.

Propagation of chaos may thus be summarized loosely by saying that, if the n-particle velocity distribution ν_t^n approximates the infinite particle distribution at $t = 0$, then the approximation persists for all time. Uchiyama [25] gives a result for the space-dependent B.E. on approximation by n-particle systems under more stringent hypotheses. We also show that Tanaka's "trend to equilibrium" results [20] remain valid under (1.3).

The results of the last few paragraphs should be compared with the last sentence of §3 in Kac [11].

Finally we note that, if $Q(d\theta)$ is replaced by $Q(u,v,d\theta)$ satisfying

$$(1.5) \qquad \int_0^{\pi} \theta^2 Q(u,v,d\theta) \leq c < \infty$$

and some mild additional conditions, then the existence results remain valid, but for uniqueness we need strong restrictions that eliminate all cases of (presumed) physical interest (we take this to mean at least inverse power laws; cf. [24], ch. XII). Funaki [7] deals with $Q(u,v,d\theta) = k(u,v)Q(d\theta)$, $\int \theta dQ < \infty$, but his conditions also exclude the physically interesting cases. Sznitman's results [19] also apply to $k(u,v)Q(d\theta)$, for Q finite.

The second author would like to thank the Indian Statistical Institute, Delhi Centre, and its faculty for their warm hospitality, and the American Institute for Indian Studies, New Delhi, for its help on a variety of matters, during the term of his Indo-American Fellowship (January-June 1988) and again in January 1989, when most of the work on this paper was completed.

2. THE n-PARTICLE PROCESS

Let $S = \mathbf{R}^3$. We write u, v, etc., for points in S, and $\underline{v} = (v_1,\ldots,v_n)$ for points in S^n. Given $u,v \in S$, $u \neq v$, $S_{u,v}$ denotes the (surface of the) sphere in $\mathbf{R}^3$ with center $u^* = (u+v)/2$ and radius $|u-v|/2$ ($|\cdot|$ denotes the Euclidean norm of the appropriate dimension). Note that $u,v \in S_{u,v}$. Regarding u as the north pole, we assign *angular co-ordinates* to any point $w \in S_{u,v}$ as follows: The *colatitude* θ, $0 < \theta \leq \pi$, satisfies $\cos\theta = 4(u-u^*)\cdot(w-u^*)/|u-v|^2$, where $\cdot$ is the usual inner produce in $\mathbf{R}^3$; it is the angle between w and the north pole. The *azimuth angle* ϕ gives the longitude of w relative to an arbitrarily chosen longitudinal "line" designated as $\phi = 0$. Thus $0 \leq \phi < 2\pi$. We write $w = (u,v,\theta,\phi)$ to denote a point w on $S_{u,v}$ with co-ordinates θ,ϕ. The choice of $\phi = 0$ on each $S_{u,v}$ is arbitrary: we only require that the function $\alpha(u,v,\theta,\phi)$ be measurable.

Let u, $v(u',v')$ be the pre- (post-) collision velocities of two particles in a binary elastic encounter. The laws of physics governing such encounters are conservation of momentum and energy:

$$(2.1) \qquad u + v = u' + v'$$

$$(2.2) \qquad |u|^2 + |v|^2 = |u'|^2 + |v'|^2 .$$

It follows that $u',v' \in S_{u,v}$ and, if θ,ϕ are the angular coordinates of u', then

$$(2.3) \qquad u' = \alpha(u,v,\theta,\phi), \qquad v' = \alpha(u,v,\pi-\theta,\phi+\pi).$$

When no confusion can arise, we use the notation u', v' as in (2.3).

We also let

$$(2.4) \qquad a(u,v,\theta,\phi) = \alpha(u,v,\theta,\phi) - u \ (= u' - u)$$

and then have

$$(2.5) \qquad |a(u,v,\theta,\phi)|^2 = |u-v|^2 \sin^2\theta/2$$

$$(2.6) \qquad \frac{1}{2\pi}\int_0^{2\pi} a(u,v,\theta,\phi)\,d\phi = u'' - u = -(u-v)\sin^2\theta/2$$

where u'' is the projection of u' on the axis of $S_{u,v}$. These relations follow from simple geometry.

The function $a(u,v,\theta,\phi)$ need not be continuous in (u,v), but

Tanaka [20] has shown that

$$(2.7)\qquad |a(u_1,v_1,\theta,\phi) - a(u_2,v_2,\theta,\phi + \phi_0(u_1,v_1,u_2,v_2))| \le c_1\theta(|u_1 - u_2| + |v_1 - v_2|)$$

for a constant c_1 and a Borel function ϕ_0 on S^4.

Let $Q(d\theta)$ be a positive measure on $(0,\pi]$ such that

$$(2.8)\qquad c_0 := \int_0^\pi \sin^2(\theta/2)Q(d\theta) < \infty.$$

Thus Q is σ-finite and $Q((\varepsilon,\pi]) < \infty$ for all $\varepsilon > 0$. The physics of the system is contained in Q; see Tanaka [20], Truesdell and Muncaster [24], ch. XII, and the discussion in §1.

The evolution of the n-particle (velocity) process $V^n(t) = (V_1^n(t),\dots,V_n^n(t))$ in S^n is described heuristically as follows: during the time interval $(t,t + dt]$, the ith and j^{th} particles collide, giving post-collision velocities $V_i^n(t+dt) = \alpha(V_i^n(t),V_j^n(t)\theta,\phi)$, $V_j^n(t+dt) = \alpha(V_i^n(t),V_j^n(t),\pi - \theta,\pi + \phi)$, with all other velocities left unchanged. The number of such collisions, having angular coordinates in $d\theta d\phi$, has intensity $\frac{1}{2\pi n} d\phi Q(d\theta)dt$. If Q is a finite measure, $V^n(t)$ can be constructed directly from a Poisson random measure. When $Q((0,\pi]) = \infty$, V^n, if it exists, will be a pure jump Markov process with all states instantaneous. We study it via the associated martingale problem and the corresponding SDE.

PRELIMINARIES

We introduce the following notation. For $\underline{u} = (u_1,\dots,u_n) \in S^n$, $0 < \theta \le \pi$, $0 \le \phi < 2\pi$, put

$$(2.9)\qquad a_{ijk}(\underline{u},\theta,\phi) := (\delta_{ik} - \delta_{jk})a(u_i,u_j,\theta,\phi)$$

for $1 \le i < j \le n$, $1 \le k \le n$, ($\delta_{ik} = 1$ if $i = k$, $= 0$ if $i \ne k$) and

(2.10) $\quad A_{ij}(\underline{u},\theta,\phi) := (a_{ij1}(\underline{u},\theta,\phi),\ldots,a_{ijn}(\underline{u},\theta,\phi))$.

Thus, if the pre-collision vector of velocities is $\underline{u}$ and the ith and j^{th} particles collide, with angular coordinates θ,ϕ, then the post-collision vector is $\underline{u}'_{ij} = \underline{u} + A_{ij}(\underline{u},\theta,\phi)$.

Let $C_b^2(\mathbb{R}^m)$ $(C_c^2(\mathbb{R}^m)$ be the space of twice continuously differentiable functions on $\mathbb{R}^m$ with bounded derivatives up through second order (with compact support). For $f \in C_b^2(\mathbb{R}^m)$, let

$$(2.11) \qquad |||f||| = \left\{\|f\|^2 + \sum_i \left\|\frac{\partial f}{\partial x_i}\right\|^2 + \sum_{i,j} \left\|\frac{\partial^2 f}{\partial x_i \partial x_j}\right\|^2\right\}^{1/2},$$

where $\|\cdot\|$ is the usual "sup" norm. ($|||\cdot|||$ depends on the dimension.)

For $f \in C_b^2(S^n)$, define

$$(2.12) \qquad K_n f(\underline{u}) = \frac{1}{2\pi n} \sum_{1\le i<j\le n} \int_0^\pi\int_0^{2\pi} (f(\underline{u}+A_{ij}(\underline{u},\theta,\phi)) - f(\underline{u}))d\phi Q(d\theta).$$

We shall prove that $K_n f$ is a continuous function.

Define a kernel $k_n(\underline{u},d\underline{y})$ on $S^n \times \mathcal{B}(S^n)$ by

$$(2.13) \qquad \int f(\underline{y})k_n(\underline{u},d\underline{y}) =$$

$$\frac{1}{2\pi n} \sum_{1\le i<j\le n} \int_0^\pi\int_0^{2\pi} f(A_{ij}(\underline{u},\theta,\phi))I_{\{A_{ij}(\underline{u},\theta,\phi)\ne 0\}}d\phi Q(d\theta).$$

Since

$$(2.14) \qquad \int |\underline{y}|^2 k_n(\underline{u},dy) = \frac{2}{2\pi n}\sum_{i<j}\int_0^\pi\int_0^{2\pi} |a(u_i,u_j,\theta,\phi)|^2 d\phi Q(d\theta)$$

$$= \frac{2c_0}{n}\sum_{i<j} |u_i - u_j|^2$$

$$= 2c_0(|\underline{u}|^2 - n^{-1}|\Sigma u_i|^2)$$

$$\le 2c_0|\underline{u}|^2,$$

it follows from App. A that, for $f \in C_b^2(S^n)$,

(2.15) $\quad M_n f(\underline{u}) = \int (f(\underline{u} + \underline{y}) - f(\underline{u}) - \nabla f(\underline{u}) \cdot \underline{y}) k_n(\underline{u}, d\underline{y})$

is well-defined and $|M_n f(\underline{u})| \leq 2c_0 \|f\| \, |\underline{u}|^2$. Let

(2.16) $\quad D_n f(\underline{u}) = -\frac{c_0}{n} \sum_{i=1}^{n} \nabla_i f(\underline{u}) \cdot (u_i - \bar{u}),$

where $\bar{u} := \frac{1}{n} \sum_1^n u_i$ and ∇_i is the gradient in $\mathbf{R}^3$ acting on the variable u_i.

LEMMA 2.1. *For* $f \in C_b^2(S^n)$, $K_n f$ *is well-defined and*

(2.17) $\quad K_n f = M_n f + D_n f$

(2.18) $\quad |K_n f(\underline{u})| \leq 4c_0 \|f\| \, (|\underline{u}|^2 + 1).$

PROOF. Observe that (2.6) implies

$$\frac{1}{2\pi} \int_0^{\pi} \int_0^{2\pi} a_{ijk}(\underline{u}, \theta, \phi) \, d\phi \, Q(d\theta) = -c_0(\delta_{ik} - \delta_{jk})(u_i - u_j),$$

hence

$$\frac{1}{2\pi n} \sum_{i<j} \int_0^{\pi} \int_0^{2\pi} \sum_k \nabla_k f(u) \cdot a_{ijk}(\underline{u}, \theta, \phi) \, d\phi \, Q(d\theta) = D_n f(\underline{u}).$$

Adding and subtracting the gradient terms in (2.12), one gets (2.17). Inequality (2.18) follows from the estimate on $M_n f$ and an obvious one on $D_n f$.

For $\varepsilon > 0$, let $Q^{\varepsilon}(B) = Q(B \cap (\varepsilon, \pi])$ and let K_n^{ε} be defined by (2.12) with Q^{ε} replacing Q; (2.7) shows that $K_n^{\varepsilon} f$ is continuous for $f \in C_c^2(S^n)$.

LEMMA 2.2. *For* $f \in C_c^2(S^n)$, $K_n^{\varepsilon} f \to K_n f$ *uniformly as* $\varepsilon \downarrow 0$, *and* $K_n f \in C_c(S^n)$.

PROOF. Since $K_n - K_n^{\varepsilon}$ is of the type (2.12) with measure $Q - Q^{\varepsilon}$, one gets

$$(2.19) \qquad |K_n f(\underline{u}) - K_n^{\varepsilon} f(\underline{u})| \le 4|||f|||\,(|\underline{u}|^2 + 1)\int_0^{\varepsilon} \sin^2 \frac{\theta}{2}\, Q(d\theta).$$

Noting that $|\underline{u} + A_{ij}(\underline{u},\theta,\phi)| = |\underline{u}|$, (2.12) shows that the supports of $K_n f$, $K_n^{\varepsilon} f$, for all $\varepsilon > 0$, are contained in a common ball, which, with (2.19), yields the lemma.

THE MARTINGALE PROBLEM.

The problem is to find a process $V^n(t)$ with paths in $\Omega_n := D_{S^n}[0,\infty)$, adapted to a filtration F^n on a probability space (Ω,F,P) such that, for $f \in C_c^2(S^n)$,

$$f(V^n(t)) - f(V^n(0)) - \int_0^t K_n f(V^n(s))ds$$

is a martingale. This is the so-called (in [EK], p.186) $D_E[0,\infty)$ *martingale problem* (here $E = S^n$) *for* K_n. Since we shall only consider solutions with cadlag trajectories, we refer simply to the K_n-martingale problem. If an initial distribution $\nu^n \in P(S^n)$ is specified, it is the (K_n,ν^n)-martingale problem. If V^n is a solution with law P^n on Ω_n, we write $V^n \in (K_n)$ or $P^n \in (K_n)$; similarly $V^n \in (K_n,\nu^n)$, $P^n \in (K_n,\nu^n)$. Here is the main result of §2.

THEOREM 2.3. *For each* $\nu^n \in P(S^n)$, *there exists a unique* $P^n \in (K_n,\nu^n)$.

This will be proved in several steps, the first being:

LEMMA 2.4. *If* $V^n \in (K^n)$, *then*, a.s., $|V^n(t)| = |V^n(0)|$ *for all* $t \ge 0$.

PROOF. Let $f(\underline{u}) = |\underline{u}|^2$ and $g \in C_c^2(\mathbf{R})$. Then $g \circ f$ and $g^2 \circ f$ are in $C_c^2(S^n)$, hence $Y_t := g \circ f(V^n(t))$ and Y_t^2 are martingales, and $Z_t := |Y_t - Y_0|^2$ is a nonnegative martingale with $Z_0 = 0$. Thus

$Z_t = 0$ for all $t \geq 0$, a.s.

Since Q^ε is a finite measure, K_n^ε is the generator of a pure-jump Markov process with no instantaneous states ([EK], ch. 4, §2). Thus, for $\nu^n \in P(S^n)$, there exists a $V^{n,\varepsilon} \in (K_n^\varepsilon, \nu^n)$. Now, given $\delta > 0$ and choosing $C = \{\underline{u}: |\underline{u}| \leq \alpha\}$ so that $\nu^n(C) > 1 - \delta$, Lemma 2.4 gives

$$\inf_{\varepsilon>0} P\{V^{n,\varepsilon}(t) \in C \text{ for } 0 \leq t \leq T\} = \nu^n(C) > 1 - \delta$$

for each T. This, Lemma 2.2, and [EK], ch. 4, 5.2, imply that $\{V^{n,\varepsilon_k}: k \geq 1\}$ is relatively compact if $\varepsilon_k \downarrow 0$; and, if V is any subsequential limit (in law) of V^{n,ε_k}, then (ibid. 5.1) $V \in (K_n, \nu^n)$. Thus the existence part of Theorem 2.3 is proven.

We first consider uniqueness for the $(K_n, \delta_{\underline{v}})$-martingale problem, $\underline{v} \in S^n$. From the equivalence between martingale problems and weak solutions to SDEs in App. A, it suffices to show that eq. (2.20) below has a unique weak solution in the class of processes V^n satisfying $E\int_0^T |V^n(t)|^2 dt < \infty$, because any solution V^n to (2.20) satisfies $|V^n(t)|^2 = |\underline{v}|^2$:

$$V^n(t) = \underline{v} + \sum_{1\leq i<j\leq n} \int_{(0,t]\times A} A_{ij}(V^n(s-)\theta,\phi)\tilde{N}_{ij}(ds\,d\theta\,d\phi) - c_0\int_0^t (V^n(s) - \bar{\bar{V}}^n(s))ds, \tag{2.20}$$

where, for $\underline{u} \in S^n$, $\bar{\bar{u}} = (\bar{u},\ldots,\bar{u}) \in S^n$ (cf. (2.16)), $A = (0,\pi]\times[0,2\pi)$, $\{N_{ij}: 1 \leq i < j \leq n\}$ is a family of independent Poisson random measures on $\mathbf{R}_+ \times A$ with respect to a filtration (G_t), with common intensity $\lambda_n(ds\,d\theta\,d\phi) = (2\pi n)^{-1} ds\,Q(d\theta)d\phi$, and $\tilde{N}_{ij} = N_{ij} - \lambda_n$ is the compensated version of N_{ij}.

We will only outline the proof of uniqueness, which follows the arguments in Tanaka [20], [22]. Let V^n be as in (2.20). We show that $L(V^n)$ is uniquely determined by $\underline{v}$, n, and Q.

For a partition Δ: $0 = t_0 < t_1 < \ldots$ of $[0,\infty)$, write $|\Delta| = \sup_i |t_{i+1} - t_i|$. The "$\Delta$-approximation" $Y = Y^\Delta$ to V^n is defined, along with some auxiliary "twist" processes ψ_{ij}, inductively, as follows. Let $Y(0) = V^n(0)$ and $\psi_{ij}(0) = 0$; if $Y(t)$, $\psi_{ij}(t)$ are defined for $0 \leq t \leq t_k$, then, for $t_k < t \leq t_{k+1}$, define

$$\psi_{ij}(t) = \phi_0(V_i^n(t-),V_j^n(t-),Y_i(t_k),Y_j(t_k)) \tag{2.21}$$

$$Y(t) = Y(t_k) + \sum_{i<j} \int_{(t_k,t]\times A} A_{ij}(Y(t_k),\theta,\phi+\psi_{ij}(s))\tilde{N}_{ij}(dsd\theta d\phi) - c_0(t - t_k)(Y(t_k) - \bar{\bar{Y}}(t_k)). \tag{2.22}$$

Observe that ψ_{ij}, $1 \leq i < j \leq n$, are (G_t)-predictable processes. Now, if N_{ij}^* is defined by

$$N_{ij}^*(B) = \int I_B(s,\theta,\phi + \psi_{ij}(s))N_{ij}(dsd\theta d\phi),$$

then the N_{ij}^* are independent Poisson measures and (2.22) holds with ψ_{ij} replaced by 0 and N_{ij} by N_{ij}^*. Thus the law of Y depends only on Δ, n, Q, and $\underline{v}$. Further, using (2.7) and Gronwall's inequality, one gets

$$E \sup_{0\leq t\leq T} |V^n(t) - Y(t)|^2 \leq C_T E\int_0^T |V^n(s) - V^n(\delta(s))|^2 ds, \tag{2.23}$$

where $\delta(s) = t_k$ on $(t_k,t_{k+1}]$, $\delta(0) = 0$, and $C_T < \infty$ depends only on c_0, c_1 (see (2.8), (2.7)) and T. Since V^n is cadlag and $|V^n(s)| = |V^n(\delta(s))| = |\underline{v}|$ by 2.4, the right member of (2.23) goes to zero as $|\Delta| \to 0$, and so

$$\lim_{|\Delta|\to 0} E \sup_{0\leq t\leq T} |V^n(t) - Y^\Delta(t)|^2 = 0. \tag{2.24}$$

Since the law of Y^Δ is uniquely determined by Q, Δ, n, it follows from (2.24) that $Y^\Delta \xrightarrow{D} V^n$ and the law of V^n is uniquely determined. Thus the SDE (2.20) has a unique weak solution, and, by Theorem A.1, we get uniqueness of the solution to the $(K_n, \delta_{\underline{v}})$-martingale problem. We denote it by $P_{\underline{v}}$.

Here and later we will need the following observation.

REMARK 2.1. In [EK], ch. 4, Theorem 4.6, suppose that the $D_E[0,\infty)$-martingale problem for (A,δ_x) is well-posed for each $x \in E$, the solution being denoted by P_x, but nothing is assumed about the well-posedness of the (A,μ)-martingale problem for general $\mu \in P(E)$. We also assume, as in [EK], that, $A \subset \bar{C}(E) \times B(E)$ and that A is contained in the bp-closure of a countable subset A_0 of A (we say that A is *separable* in this case). Then (i) *The map* $x \longrightarrow P_x$ *is measurable*, and (ii) *The* (A,μ)*-martingale problem is well-posed, the unique solution being* $P_\mu = \int P_x \mu(dx)$.

Noticing that the set $M = \{\delta_x \colon x \in E\}$ is a Borel (in fact,closed) set in $P(E)$, (i) follows as in [EK]. Concerning (ii), it is easy to check that $P_\mu \in (A,\mu)$. Suppose $P \in (A,\mu)$ and that $Q_{\omega'}$ is a regular version of $P(\cdot|\sigma(X_0))$. The argument of [18], Theorem 1.2.10, yields $Q_{\omega'} = P_{X_0(\omega')}$, which in turn gives $P = P_\mu$.

Finally note that, if there exists a countable set $\mathcal{D}_0 \subset \mathcal{D}(A)$ such that, for $f \in \mathcal{D}(A)$, there exist $f_k \in \mathcal{D}_0$ with $\|f_k - f\| \to 0$ and $\|Af_k - Af\| \to 0$, then A is separable.

These observations, (2.18), and the separability of $C_c^2(S^n)$ in the $|||\cdot|||$-norm (see [EK], App. 7), imply that the (K_n,ν^n)-martingale problem is well-posed, with unique solution

$$P = \int P_{\underline{v}}\, \nu^n(d\underline{v}). \tag{2.26}$$

This completes the proof of Theorem 2.3.

REMARK 2.2. The estimate (2.18) also holds for $f \in C_b^2(S^n)$, thus

$f(V^n(t)) - f(V^n(0)) - \int_0^t K_n f(V^n(s))ds$ is a local martingale for $f \in C_b^2(S^n)$.

Theorem 2.3 implies that $\{P_{\underline{v}}: \underline{v} \in S^n\}$ is a strong Markov process with a Feller semigroup ([EK], ch. 4, 4.2 and 5.1).

ESTIMATES ON V^n.

These will be used in §3. We regard V^n as a solution of the martingale problem or of the SDE, depending on convenience.

THEOREM 2.5. *Let* $V^n \in (K_n, \nu^n)$. *There exist constants* C_1, C_2 *such that*

$$(2.27) \qquad E \sum_{i=1}^n |V_i^n(t)|^4 \le C_1 e^{C_2 t} E \sum_{i=1}^n |V_i^n(0)|^4 \quad .$$

PROOF. Let $f(\underline{u}) = \sum_1^n |u_i|^4$. Although $f \notin C_b^2(S^n)$, $K_n f$ can be defined, e.g., by (2.17).

Fix $\underline{v} \in S^n$ and $V^n \in (K_n, \delta_{\underline{v}})$. By 2.4, $|V^n(t)|^2 = |\underline{v}|^2$. Let $g \in C_c^2(S^n)$ be such that $f(\underline{u}) = g(\underline{u})$ on $\{|\underline{u}|: |\underline{u}|^2 \le 2|\underline{v}|^2\}$. Observe that $K_n f(\underline{u}) = K_n g(\underline{u})$ for $|\underline{u}| \le |\underline{v}|$, so

$$f(V^n(t)) - f(V^n(0)) - \int_0^t K_n f(V^n(s))ds$$

is a martingale. Fix i, j, and $\underline{u} \in S^n$, and write $u_i' = \alpha(u_i, u_j, \theta, \phi)$, $u_j' = \alpha(u_i, u_j, \pi-\theta, \pi+\phi)$; then

$$\int(|u_i'|^4 - |u_i|^4)d\phi = \int[(|(u_i' - u_i) + u_i|^2)^2 - |u_i|^4]d\phi$$

$$= \int[(|u_i'-u_i|^2 + |u_i|^2 + 2(u_i'-u_i)\cdot u_i)^2 - |u_i|^4]d\phi.$$

By (2.5) and (2.6) we find, for some c_3,

$$|\int(|u_i'|^4 - |u_i|^4)d\phi| \leq c_3 \sin^2\left(\frac{\theta}{2}\right)(|u_i|^4 + |u_j|^4),$$

with a similar inequality for $\int(|u_j'|^4 - |u_j|^4)d\phi$. Summing over i,j and integrating wrt $Q(d\theta)$ one gets

$$|K_n f(\underline{u})| \leq c_4 f(\underline{u}).$$

Thus

$$\begin{aligned} Ef(V^n(t)) &= Ef(V^n(0)) + E\int_0^t K_n f(V^n(s))ds \\ &\leq Ef(V^n(0)) + c_4\int_0^t Ef(V^n(s))ds, \end{aligned}$$

and (2.27) follows by Gronwall's inequality since, from 2.4,

$$Ef(V^n(t)) \leq E\left(\sum_1^n |V_i^n(t)|^2\right)^2 = |\underline{v}|^4 .$$

The general case, $V^n \in (K_n, \nu^n)$, follows by (2.26).

The next inequality is standard for SDEs with Lipschitz coefficients. Here we prove it for weak solutions via the Δ-approximation.

THEOREM 2.6. *Let* $V^n, W^n \in (K_n)$, $V^n(0) = \underline{v}$, $W^n(0) = \underline{w}$. *Then, on some probability space, there exist processes* $\hat{V}^n \stackrel{D}{=} V^n$ *and* $\hat{W}^n \stackrel{D}{=} W^n$ *such that*

$$\text{(2.28)} \qquad E \sup_{t\leq T} |\hat{V}^n(t) - \hat{W}^n(t)|^2 \leq C_T|\underline{v} - \underline{w}|^2 ,$$

where the constant C_T *depends only on* Q *and* T .

PROOF. Let N_{ij} be as in (2.20). Define a "Δ-approximation" Y^Δ of V^n by (2.22) with $Y^\Delta(0) = \underline{v}$, $\psi_{ij} \equiv 0$, and a "Δ-approximation" Z^Δ of W^n by (2.22) with $Z^\Delta(0) = \underline{w}$ and

$$\psi_{ij}(s) = \phi_0(Y_i^\Delta(t_k), Y_j^\Delta(t_k), Z_i^\Delta(t_k), Z_j^\Delta(t_k)), \; t_k < s \leq t_{k+1} .$$

Using (2.7) and Gronwall, we get

$$\text{(2.29)} \qquad E \sup_{t\leq T} |Y^\Delta(t) - Z^\Delta(t)|^2 \leq C_T|\underline{v} - \underline{w}|^2 .$$

As $|\Delta| \to 0$, $Y^{\Delta} \xrightarrow{D} V^n$ and $Z^{\Delta} \xrightarrow{D} W^n$ by the remarks after (2.22) and (2.24). Thus, if $|\Delta_k| \to 0$, there is a subsequence, also denoted by Δ_k, such that $(Y^{\Delta_k}, Z^{\Delta_k}) \xrightarrow{D} (\hat{V}^n, \hat{W}^n)$ (say). Now Fatou's lemma and (2.29) yield (2.28).

We conclude this section with bounds on the modulus of continuity of V^n, to be used in proving tightness in the next section.

THEOREM 2.7. *Suppose* $V^n \in (K_n)$ *with* $E|V^n(0)|^2 < \infty$. *For* $T > 0$ *there exist constants* C_T, C_T' *such that, for* $G_t = \sigma(V^n(s): s \le t)$,

$$(2.30) \qquad E\left[\sup_{t_1 \le t \le t_1+T} |V_1^n(t)|^2 | G_{t_1}\right] \le C_T\left(|V_1^n(t_1)|^2 + \frac{1}{n}|V^n(0)|^2\right)$$

and, for $t_1 \le t \le t_1 + T$,

$$(2.31) \qquad E\left[|V_1^n(t)-V_1^n(t_1)|^2 | G_{t_1}\right] \le C_T'(t-t_1)\left(|V_1^n(t_1)|^2 + \frac{2}{n}|V^n(0)|^2\right)$$

$$(2.32) \qquad E\left[|V^n(t)-V^n(t_1)|^2 | G_{t_1}\right] \le 3C_T'(t-t_1)|V^n(0)|^2.$$

PROOF. It suffices to consider $t_1 = 0$ and $V^n(0) = \underline{v}$, so that conditional expectations become unconditional. From (2.20),

$$V_1^n(t) = v_1 + \sum_{j\ne 1} \int_{(0,t]\times A} a(V_1^n(s-), V_j^n(s-)\theta, \phi)\tilde{N}_{1j}(ds\,d\theta\,d\phi)$$
$$- c_0\int_0^t (V_1^n(s) - \bar{V}^n(s))ds.$$

Thus

$$E\sup_{s\le t}|V_1^n(s)|^2 \le 3|v_1|^2 + 6E\sum_{j\ne 1}\frac{c_0}{n}\int_0^t (|V_1^n(s)|^2 + |V_j^n(s)|^2)ds$$
$$+ 6c_0^2 tE\int_0^t (|V_1^n(s)|^2 + |\bar{V}^n(s)|^2)ds$$
$$\le 3|v_1|^2 + 6c_0(1+c_0t)E\int_0^t \left(|V_1^n(s)|^2 + \frac{1}{n}\sum_{j=1}^n |V_j^n(s)|^2\right)ds,$$

and Lemma 2.4 and Gronwall's inequality yield (2.30).

For (2.31), similar arguments give

$$E|V_1^n(t) - v_1|^2 \leq 4c_0(1+c_0t)\left(E\int_0^t |V_1^n(s)|^2 ds + \frac{t}{n}\,|\underline{v}|^2\right) ,$$

and (2.30) implies

$$E|V_1^n(t) - v_1|^2 \leq 4tc_0C_T(1+c_0t)(|v_1|^2 + \tfrac{2}{n}|\underline{v}|^2) .$$

Estimates like (2.31) for the other components are now added up to yield (2.32).

Recalling K_n^ε from Lemma 2.2, we have the following corollary.

COROLLARY 2.8. *Let* $V^n \in (K_n,\mu)$, $V^{n,\varepsilon} \in (K_n^\varepsilon,\mu)$. *Then, on a suitable probability space, there exist processes* $W^n \overset{D}{=} V^n$ *and* $W^{n,\varepsilon} \overset{D}{=} V^{n,\varepsilon}$ *such that*

(2.33) $$E \sup_{t\leq T} |W_1^n(t) - W_1^{n,\varepsilon}(t)|^2 \leq C_T E|V_1^n(0)|^2 \int_0^\varepsilon \theta^2 Q(d\theta),$$

(2.34) $$E \sup_{t\leq T} |W^n(t) - W^{n,\varepsilon}(t)|^2 \leq C_T \int |v|^2 \mu(dv) \int_0^\varepsilon \theta^2 Q(d\theta)$$

for a constant C_T *depending only on* T *and* Q.

3. BOLTZMANN PROCESSES

Let $V^n(t)$ be as in §2, and define $U_t^n := V_1^n(t)$, $\nu_t^n := n^{-1}\sum_1^n \delta_{V_i^n(t)}$, and $\beta_t^n := (U_t^n, \nu_t^n)$; β^n is the *n-particle Boltzmann process*. Here we study the behavior of β^n as $n \to \infty$, specifically tightness and partial identification of the limit (which is completed in §4).

Let $P_2(S)$ be the set of $\mu \in P(S)$ such that $b^2(\mu) := \int |v|^2 \mu(dv)$ is finite. For $\mu_1, \mu_2 \in P_2(S)$,

(3.1) $\rho_2(\mu_1,\mu_2)$

$$:= \inf\{(\int|v_1-v_2|^2F(dv_1dv_2))^{\frac{1}{2}}: F \in P(S^2) \text{ with marginals } \mu_1,\mu_2\}$$

defines a metric on $P_2(S)$, and $\rho_2(\mu_k,\mu) \to 0$ iff $\mu_k \xrightarrow{w} \mu$ and $b^2(\mu_k) \to b^2(\mu)$. If $f \in C(S)$ and $|f(v)| \leq c|v|^2$, then $\rho(\mu_k,\mu) \to 0$ implies $\langle\mu_k,f\rangle \to \langle\mu,f\rangle$. Also, if $\mu_1 = \frac{1}{m}\sum_1^m \delta_{v_i}$, $\mu_2 = \frac{1}{m}\sum_1^m \delta_{w_i}$, then

$$(3.2) \qquad \rho_2(\mu_1,\mu_2) \leq \left(\frac{1}{m}\sum_1^m |v_i - w_i|^2\right)^{\frac{1}{2}} .$$

The following sets are compact in $P_2(S)$:

$$\{\mu: \int|v|^{2+\delta}\mu(dv) \leq c\}, \qquad \delta > 0, \qquad c < \infty .$$

Let $E = S \times P_2(S)$. For $n \geq 1$ define $R_n: S^n \to E$ by

$$(3.3) \qquad R_n(v_1,\dots,v_n) = \left(v_1, \frac{1}{n}\sum_1^n \delta_{v_i}\right) ,$$

E_n = range of R_n, so $\beta_t^n = R_n(V^n(t))$.

THEOREM 3.1. *Let* $V^n \in (K_n)$; *then* β^n *is a Markov process. Further, let* $\bar{P}_z^n := P_{\underline{v}}^n \circ (\beta^n)^{-1}$, *where* $\underline{v}$ *satisfies* $R_n(\underline{v}) = z$ *and* $P_{\underline{v}}^n \in (K_n,\delta_{\underline{v}})$. *Then*

$$(3.4) \qquad P(\beta^n \in B|\beta_0^n) = \bar{P}_{\beta_0^n}^n(B), \quad B \in \mathcal{B}(D_E[0,\infty)).$$

PROOF. We show first that $\bar{P}_z^n$ is well-defined, i.e., if $R_n(\underline{v}) = R_n(\underline{w})$, then $P_{\underline{v}}^n \circ (\beta^n)^{-1} = P_{\underline{w}}^n \circ (\beta^n)^{-1}$. Let V^n be the coordinate process on $D_{S^n}[0,\infty)$. Since $K_n(f \circ \sigma) = K_n f \circ \sigma$ for any permutation σ of $\{1,\dots,n\}$, it follows that, under $P_{\underline{v}}^n$, $\sigma V^n \in (K_n,\delta_{\sigma\underline{v}})$, where $\sigma(v_1,\dots,v_n) = (v_{\sigma 1},\dots,v_{\sigma n})$.

Now $R_n(\underline{v}) = R_n(\underline{w})$ means $\underline{w} = \sigma\underline{v}$ for some σ such that $\sigma 1 = 1$. Thus $\tilde{V}^n := \sigma V^n \in (K_n,\delta_{\underline{w}})$ and so

$$\begin{aligned} P^n_{\underline{v}}((\beta^n)^{-1}B) &= P^n_{\underline{v}}(R_n(V^n) \in B) \\ &= P^n_{\underline{v}}(R_n(\tilde{V}^n) \in B) \\ &= P^n_{\underline{v}}(\tilde{V}^n \in R_n^{-1} \in B) \\ &= P^n_{\underline{w}}((\beta^n)^{-1}B). \end{aligned}$$

Let $F^n_t = \sigma(V^n(s): s \leq t)$ and $A \in \mathcal{B}(E)$. Then

$$\begin{aligned} P(\beta^n_{s+t} \in A | F^n_t) &= P(V^n(s+t) \in R_n^{-1} A | F^n_t) \\ &= P^n_{V^n(t)}(V^n(s) \in R_n^{-1} A) \\ &= \bar{P}^n_{\beta^n_t}(R_n(V^n(s)) \in A). \end{aligned}$$

This completes the proof.

REMARK 3.1. Similarly, the second component of β^n is itself a Markov process.

We regard $\bar{P}^n_z$, $z \in E_n$, as a measure on $D_E[0,\infty)$.

THEOREM 3.2. *Let* $z_n = (u^n, \nu^n) \in E_n$ *such that* $z_n \to z = (u,\nu) \in E$ (convergence in E); *then* $\{\bar{P}^n_{z_n}\}$ *is relatively compact.*

PROOF. 1°. Let $\underline{v}^n = (v^n_1,\ldots,v^n_n) \in S^n$, $R_n(\underline{v}^n) = z_n$. Thus

$$v^n_1 \to u \tag{3.5}$$

$$\frac{1}{n}\sum_{i=1}^{n} |v^n_i|^2 \to \int |v|^2 \nu(dv), \tag{3.6}$$

hence, for some constant $c' < \infty$.

$$|v^n_1|^2 + \frac{1}{n}|\underline{v}^n|^2 \leq c' \quad . \tag{3.7}$$

Let $V^n \in (K_n, \delta_{\underline{v}^n})$; then $\bar{P}^n_{z_n} = L(\beta^n)$. It suffices to show separately that $\{L(V^n_1)\}$ and $\{L(\nu^n_\cdot)\}$ are tight. For each of these we prove that the "containment condition" in [EK], ch. 3, Theorem 7.2(a), holds and an estimate on the conditional expected increments (ibid., Remark 8.7); then [EK], ch. 3, Theorem 8.6, yields tightness.

2°. *The laws* $\{L(V_1^n)\}$ *are tight.*

Using (2.30), with $t_1 = 0$, and (3.7), we get

$$(3.8)\qquad E\left[\sup_{0\le t\le T} |V_1^n(t)|^2\right] \le C_T C',$$

and, by (2.31), for $t_1 \le t \le t_1 + \delta$, $t_1, t \le T$, and Lemma 2.4,

$$(3.9)\qquad E\left[|V_1^n(t) - V_1^n(t_1)|^2 | G_{t_1}^n\right] \le \delta C_T'\left\{\sup_{0\le t\le T} |V_1^n(t)|^2 + 2c'\right\},$$

where $G_{t_1}^n = \sigma(V^n(s): s \le t)$. The result follows.

3°. *The laws* $\{L(\nu_\cdot^n)\}$ *are tight.*

First we bound the increments. By (3.2),

$$\rho_2^2(\nu_t^n, \nu_{t_1}^n) \le \frac{1}{n} \sum_{i=1}^n |V_i^n(t) - V_i^n(t_1)|^2$$
$$= \frac{1}{n} |V^n(t) - V^n(t_1)|^2 .$$

Thus (2.32) yields, for $t_1 \le t \le t_1 + \delta$, $t, t_1 \le T$,

$$(3.10)\qquad E\left[\rho_2^2(\nu_t^n, \nu_{t_1}^n) | G_{t_1}^n\right] \le 3c_T'\delta n^{-1} |V^n(t_1)|^2$$
$$\le 3c_T' c' ,$$

since $|V^n(t_1)|^2 = |\underline{v}^n|^2 \le nc'$ by Lemma 2.4.

As for the containment condition, given $\varepsilon > 0$, t fixed, we shall find a compact set $\Gamma_{\varepsilon,t}$ in $P_2(S)$ such that

$$(3.11)\qquad \limsup_{n\to\infty} P(\nu_t^n \in \Gamma_{\varepsilon,t}^\varepsilon) \ge 1 - \varepsilon,$$

where $\Gamma^{\varepsilon}_{\varepsilon,t} = \{\mu: \rho_2(\mu,\Gamma_{\varepsilon,t}) < \varepsilon\}$.

Define $G_p: S \to S$, $p > 0$, by $G_p(v) = (|v| \wedge p/|v|)v$ if $v \neq 0$ $= 0$ if $v = 0$, let $\underline{w}^n = (w_1^n,\ldots,w_n^n)$, where $w_i^n = G_p(v_i^n)$, and let $W^n \in (K_n, \delta_{\underline{w}^n})$. By Theorem 2.6 we construct $\tilde{V}^n \overset{D}{=} V^n$, $\tilde{W}^n \overset{D}{=} W^n$ such that (2.28) holds. Define

$$\tilde{\nu}_t^n = \frac{1}{n}\sum_{i=1}^n \delta_{\tilde{V}_i^n(t)}, \qquad \pi_t^n = \frac{1}{n}\sum_{i=1}^n \delta_{\tilde{W}_i^n(t)} .$$

Then $\tilde{\nu}^n \overset{D}{=} \nu^n$, and, by (2.28) and (3.2),

$$\begin{aligned} E\rho_2^2(\tilde{\nu}_t^n,\pi_t^n) &\le c_t \frac{1}{n}\sum_{i=1}^n |v_i^n - w_i^n|^2 \\ &= c_t \frac{1}{n}\sum_{i=1}^n |v_i^n - G_p(v_i^n)|^2 \\ &= c_t \int |v - G_p(v)|^2 \nu_0^n(dv). \end{aligned}$$

Thus

$$\text{(3.12)} \qquad \begin{aligned} \lim_{p\to\infty}\limsup_{n\to\infty} E\rho_2^2(\tilde{\nu}_t^n,\pi_t^n) &\le \lim_{p\to\infty} c_t \int |v - G_p(v)|^2 \nu_0(dv) \\ &= 0 , \end{aligned}$$

since $\nu_0^n \to \nu_0$ in $P_2(S)$. Thus we can choose $p < \infty$ such that

$$\text{(3.13)} \qquad \limsup_n P(\rho_2(\tilde{\nu}_t^n,\pi_t^n) > \varepsilon) < \varepsilon/2,$$

and (2.27) yields

$$\begin{aligned} E\int |v|^4 \pi_t^n(dv) &\le \frac{1}{n} C_1 e^{C_2 t} E\sum_{i=1}^n |w_i^n(0)|^4 \\ &\le C_1 e^{C_2 t} p^4 . \end{aligned}$$

Hence we can choose $q > 0$ such that

$$\text{(3.14)} \qquad \sup_n P\left(\int |v|^4 \pi_t^n(dv) > q\right) \le \varepsilon/2 .$$

Now

$$\Gamma_{\varepsilon,t} = \{\mu \in P_2(S)\colon \int |v|^4 \mu(dv) \le q\}$$

is compact, (3.14) implies

(3.15) $\quad P(\pi_t^n \in \Gamma_{\varepsilon,t}) \ge 1 - \varepsilon/2,$

and (3.11) follows from (3.13) since $\tilde{\nu}^n \overset{D}{=} \nu^n$.

THE L_n-MARTINGALE PROBLEMS AND THEIR LIMIT

The Markov process β^n is a solution to the L_n-martingale problem, L_n being a suitable restriction of the generator of β^n. We show that $L = \lim L_n$ exists and that any subsequential limit of β^n solves the L-local martingale problem. In the next section we will show that the L-local martingale problem is well-posed, and this gives convergence of β^n.

For $f,g \in C_c^2(S)$ define $H_{fg}\colon E \to \mathbf{R}$ by

(3.16) $\quad H_{fg}(u,\nu) = f(u)\exp\langle \nu, g\rangle,$

and $H_{1g}(u,\nu) = \exp\langle \nu, g\rangle$. Let

$$E_0 = \{H_{fg}\colon f \in C_c^2(S) \text{ or } f = 1,\ g \in C_c^2(S)\} .$$

Clearly $H_{fg} \circ R_n \in C_b^2(S^n)$, and, as noted earlier, $K_n(H_{fg} \circ R_n)$ is invariant under any permutation of the second through n^{th} components of points in S^n. Thus there exists a function, denoted by $L_n H_{fg}$, such that

(3.17) $\quad (L_n H_{fg}) \circ R_n = K_n(H_{fg} \circ R_n).$

Also, $V^n \in (K_n)$ implies $\beta^n = R_n(V^n) \in (L_n)_{loc}$, i.e.,

(3.18) $\quad H(\beta_t^n) - H(\beta_0^n) - \int_0^t L_n H(\beta_s^n)ds, \qquad H \in E_0 ,$

is a local martingale (cf. Remark 2.2).

For $g \in C_c^2(S)$, and recalling $A = (0,\pi] \times [0,2\pi)$, put

(3.19) $$Kg(u,v) = \frac{1}{2\pi}\int_A (g(u') - g(u))d\phi Q(d\theta),$$

where $u' = \alpha(u,v,\theta,\phi)$ as in (2.3). Also let

$$G_n(u,v,\theta,\phi) = e^{\frac{1}{n}(g(u')+g(v')-g(u)-g(v))} - 1 .$$

For $\underline{u} = (u_1,\ldots,u_n)$ and $\nu = n^{-1}\sum_1^n \delta_{u_i}$, a computation yields

(3.20) $$L_n H_{fg}(u_1,\nu) =$$

$$= e^{<\nu,g>}<\nu,Kf(u_1,\cdot)>+H_{fg}(u_1,\nu)\frac{1}{4\pi}\int_{S\times S\times A} nG_n d\phi Q(d\theta)\nu^{\otimes 2}(dudv)$$

$$+ (e^{<\nu,g>}/2\pi)\int_{S\times A}(f(u_1')-f(u_1))G_n(u_1,v,\theta,\phi)d\phi Q(d\theta)\nu(dv) .$$

To verify this, use the equation

(3.21) $$K_n(FG)(\underline{u}) =$$

$$= G(\underline{u})K_nF(\underline{u})+F(\underline{u})K_nG(\underline{u})$$

$$+ \frac{1}{2\pi n}\sum_{i<j}\int_A (F(\underline{u}')-F(\underline{u}))(G(\underline{u}')-G(\underline{u}))d\phi Q(d\theta)$$

($\underline{u}' = (u_1,\ldots,u_i',\ldots,u_j',\ldots,u_n)$ in the integral) on the functions $F(\underline{u}) = f(u_1)$, $G(\underline{u}) = e^{<\nu,g>}$.

We now *define* $L_nH_{fg}(u,\nu)$ by (3.20) for all $u \in S$, $\nu \in P_2(S)$. Taking the limit as $n \to \infty$ formally in (3.20), we define

(3.22) $$LH_{fg}(u,\nu) = e^{<\nu,g>}<\nu,Kf(u,\cdot)>+H_{fg}(u,\nu)\int_{S\times S} Kg(u,v)\nu^{\otimes 2}(dudv).$$

THEOREM 3.3. *The function* LH_{fg} *is continuous on* $E = S \times P_2(S)$. *Also,*

(3.23) $$|LH_{fg}(u,\nu)| \le C_{fg}\{|u|^2 + b^2(\nu) + 2\}$$

(3.24) $$|L_nH_{fg} - LH_{fg}| \le \frac{1}{n} C_{fg}\{|u|^2 + b^2(\nu)\}$$

(3.25) $|L_n H_{fg}(u,\nu)| \le C_{fg}\{|u|^2 + b^2(\nu) + 2\}$.

where C_{fg} *is a constant depending on* f,g *but not on* n.

PROOF. Since $Kg(u,v) = K_2(g \otimes 1)(u,v)$, where $g \otimes 1(u,v) = g(u)$, the continuity of Kg follows from Lemmas 2.1, 2.2, and we get, from (2.18),

(3.26) $|Kg(u,v)| \le 2c_0 |||g||| \, (|u|^2 + |v|^2 + |u| + |v|)$.

Together with (3.22) and properties of the ρ_2-metric, (3.26) implies that LH_{fg} is continuous on E; (3.23) also follows readily.

Subtracting (3.22) from (3.20) (with $u_1 = u$) we obtain

(3.27) $$H_{fg}(u,\nu)\frac{1}{4\pi}\int_{S\times S\times A}(nG_n-(g(u')-g(u)+g(v')-g(v)))d\phi Q(d\theta)\nu^{\otimes 2}(dudv) + III,$$

where III is the third term in (3.20). Using $|e^x - 1| \le |x|$, we have

$$|III| \le (4c_0/n)\|f\| \, \|g\| e^{\|g\|} \, (|u|^2 + b^2(\nu)).$$

Similarly, since $|e^x - 1 - x| \le x^2 e^{|x|}$, the first term in (3.27) is dominated by

$$(4c_0/n)\|f\| e^{5\|g\|} b^2(\nu) .$$

This proves (3.24), and (3.25) is then immediate.

THEOREM 3.4. *Let* P *be a subsequential limit of* $P^n := \bar{P}^n_{z_n}$ *in Theorem* 3.2; *then* $P \in (L,\delta_z)$.

PROOF. Let $P^{n_k} \xrightarrow{w} P$ on $D_E[0,\infty)$; we drop the subscript k for simplicity. By the Skorokhod representation ([EK], ch. 3, 1.8), we may assume $\beta^n \to \beta$ a.s., where $\beta_t := (U_t,\nu_t)$, $P^n = L(\beta^n)$, and

$P = L(\beta)$. By (3.25) and Lemma 2.4,

$$(3.28) \qquad |L_n H_{fg}(\beta_t^n)| \le C_{fg}\{\sup_{t\le T}|U_t^n|^2 + b^2(\nu_0^n) + 2\}, \qquad t \le T;$$

the right side of (3.28) is integrable by (3.8). Since $L_n H_{fg}(\beta_t^n) \to LH_{fg}(\beta_t)$ a.s. in view of (3.24), we have

$$(3.29) \qquad \int_0^t L_n H_{fg}(\beta_s^n)ds \to \int_0^t LH_{fg}(\beta_s)ds \quad \text{in } L^1.$$

Now (3.28) shows that (3.18), with $H = H_{fg}$, is in fact a martingale, and it follows from (3.29) that

$$H_{fg}(\beta_t) - H_{fg}(\beta_0) - \int_0^t LH_{fg}(\beta_s)ds$$

is also a martingale, i.e., $P \in (L,\delta_z)$. QED.

We will show in Theorem 4.10 that the (L,μ)-*martingale* problem is well-posed for $\mu = \delta_z$, $z \in E$; for general $\mu \in P(E)$ we must consider the (L,μ)-*local* martingale problem $((L,\mu)_{loc})$; see 4.11. Here is the main result of this section.

THEOREM 3.5. *Let* $V^n \in (K_n)$ *and* $\beta^n = R_n(V^n) \in (L_n)$, $n \ge 1$; *then* $L(\beta_0^n) \xrightarrow{D} \mu$, $\mu \in P(E)$, *implies*

$$(3.30) \qquad \beta^n \xrightarrow{D} \beta \quad \text{in} \quad D_E[0,\infty),$$

where $\beta \in (L,\mu)_{loc}$.

PROOF. As in Theorem 3.2, consider $\bar{P}_{z_n}^n$, where $z_n \to z$. By Theorems 3.4 and 4.10, $\bar{P}_{z_n}^n \to \bar{P}_z \in (L,\delta_z)$. Thus, if h is a bounded continuous function on $D_E[0,\infty)$, we have $\bar{h}_n(z_n) \to \bar{h}(z)$, where $\bar{h}_n(z_n) = \int h d\bar{P}_{z_n}^n$, $\bar{h}(z) = \int h d\bar{P}_z$.

Let $\mu^n = L(\beta_0^n)$, so $Eh(\beta^n) = \langle\mu^n,\bar{h}_n\rangle$. Now $\mu^n \xrightarrow{W} \mu$ on E, and, by the Skorokhod representation ([EK], ch. 3. 1.8), $\langle\mu^n,\bar{h}_n\rangle \to \langle\mu,\bar{h}\rangle = Eh(\beta)$, yielding (3.30). Since $\langle\mu,\bar{h}\rangle = \int h d\bar{P}_\mu$, where $\bar{P}_\mu = \int \bar{P}_z \mu(dz) \in (L,\mu)_{loc}$ by Theorem 4.11, the proof is complete.

The limit process $\beta_t = (U_t,\nu_t)$ is called the *(infinite particle) Boltzmann process.*

4. THE BOLTZMANN EQUATION AND RELATED MARTINGALE PROBLEMS

Let B_2 denote the space of measurable functions $t \longrightarrow \eta_t$ from $\mathbb{R}_+$ to $P_2(S)$ such that

(4.1) $$\sup_{t\leq T} b^2(\eta_t) < \infty \quad \textit{for each} \quad T < \infty,$$

where $b^2(\eta) = \int |v|^2 \eta(dv)$, $\eta \in P_2(S)$. Note that $D_{P_2(S)}[0,\infty) \subset B_2$.

The next lemma gives a useful bound on Kg, defined in (3.19), for $g \in C_c^2(S)$; see (3.26) for another bound.

LEMMA 4.1. *For* $g \in C_c^2(S)$, *there exists a constant* C_g *such that*

(4.3) $$|Kg(u,v)| \leq C_g(1 + |v|^2).$$

PROOF. Let $\operatorname{supp} g \subset \{v: |v| \leq M\}$, $M \geq 1$. Fix u with $|u| > 2M$ and $v \in S$, and write $u' = \alpha(u,v,\theta,\phi)$. Then $g(u) = 0$ so

(4.4) $$|Kg(u,v)| \leq \frac{1}{2\pi}\int_0^\pi\int_0^{2\pi} |g(u')|\, d\phi Q(d\theta)$$

$$\leq \frac{\|g\|}{2\pi}\int_0^\pi\int_0^{2\pi} I_{\{(\theta,\phi):|u'|\leq M\}}\, d\phi Q(d\theta).$$

If $|u| > 2M$ and $|u'| \leq M$, then $|u - u'| \geq |u| - |u'| \geq \frac{1}{2}|u|$. Thus, by (2.5), $4|u - v|^2 \sin^2\theta/2 \geq |u|^2$, which implies

$$I_{\{|u'|\leq M\}} \leq 8(1 + |v|^2)\sin^2\theta/2 \quad .$$

Thus (4.4) gives $|Kg(u,v)| \le 8c_0\|g\|(1 + |v|^2)$ for $|u| > 2M$. This and (3.26) for $|u| \le 2M$ give (4.3).

MARTINGALE PROBLEMS

Let U_t be the coordinate function on $\Omega_0 = D_S[0,\infty)$. Given $\mu_0 \in P_2(S)$ and $\eta_. \in B_2$, a probability measure P on Ω_0 is a *solution of the* $B[\eta_.,\mu_0]$*-martingale problem* if

$$(4.5) \qquad L(U_0) = \mu_0$$

and, for every $g \in C_c^2(S)$,

$$(4.6) \qquad g(U_t) - g(U_0) - \int_0^t \langle \eta_s, Kg(U_s,\cdot)\rangle ds$$

is a martingale. As an abbreviation we write $P \in B[\eta_.,\mu_0]$.

Note that (4.3) and $\eta_. \in B_2$ imply that (4.6) is bounded, hence is a local martingale iff it is a martingale.

The measure P on Ω_0 is a solution to the *Boltzmann martingale problem* with initial law μ_0, written $P \in B[\mu_0]$, if (4.5) and (4.6) hold with $\eta_t = L(U_t)$, i.e., if $P \in B[L(U_.),\mu_0]$.

REMARK 4.1. Funaki [7] refers to these as "nonlinear" martingale problems. The problem $B[\eta_.,\mu_0]$ is simply a time-inhomogeneous martingale problem as in [EK], ch. 4. The Boltzmann martingale problem $B[\mu_0]$ is not "classical", hence the standard results are not applicable.

We now turn to existence and uniqueness of solutions to the $B[\eta_.,\mu_0]$ and $B[\mu_0]$-martingale problems. First note that

$$(4.7) \quad \langle \eta_s, Kf(u,\cdot)\rangle = \int (f(u+y)-f(u)-\nabla f(u)\cdot y)k_s(u,dy) - c_0\Big(u - \int v\eta_s(dv)\Big)\cdot \nabla f(u),$$

where, for $\eta_. \in B_2$, the kernel $k_s(u,dy)$ is given by

(4.8) $\int f(y)k_s(u,dy) = \frac{1}{2\pi}\int f(a(u,v,\theta,\phi))I_{\{a(u,v,\theta,\phi)\neq 0\}}d\phi Q(d\theta)\eta_s(dv).$

Lemma (14.50) and exercise 14.4 in [J] give a measurable process $Y_s(x)$ on a suitable probability space $(X,\mathcal{X},F)$ such that $L(Y_s) = \eta_s$. Then (4.8) can be rewritten as

(4.9) $\int f(y)k_s(u,dy) = \int f(a(u,Y_s(x),\theta,\phi))I_{\{a(u,Y_s(x),\theta,\phi)\neq 0\}}d\phi Q(d\theta)F(dx).$

Here

$$\int |y|^2 k_s(u,dy) = c_0\int |u - v|^2\eta_s(dv)$$
$$\leq 2c_0(|u|^2 + b^2(\eta_s)),$$

and $\eta_. \in \mathcal{B}_2$ implies that (A.2) of App. A holds (with $z = (x,\theta,\phi)$). Thus Theorem A.1 yields

THEOREM 4.2. *Let* $\eta_. \in \mathcal{B}_2$ *and* $\mu_0 \in P_2(S)$. *Then* $P \in B[\eta_.,\mu_0]$ *iff* P *is a weak solution to the SDE*

(4.10) $$U_t = U_0 + \int_{(0,t]\times X\times A} a(U_{s-},Y_s(x),\theta,\phi)\tilde{N}(dsdxd\theta d\phi)$$
$$- c_0\int_0^t\int_S (U_s-v)\eta_s(dv)ds,$$

where N *is a Poisson measure on* $[0,\infty) \times X \times A$ *with intensity* $d\lambda_N = (2\pi)^{-1}dsF(dx)Q(d\theta)d\phi$, *and* $Y_s(x)$ *is a measurable process on* $(X,\mathcal{X},F)$ *with* $L(Y_s) = \eta_s$.

Equation (4.10) is similar to equation (3.8) of Tanaka [22], except that Tanaka uses a coordinate σ on the unit sphere in $\mathbb{R}^3$ instead of our longitudinal coordinate ϕ. (The matrix $R(s,\omega,\tilde{\omega})$ there can be omitted as per Remark 3 of [22].) Thus, translating Tanaka's argument for Proposition 2 of [22], we have

THEOREM 4.3. *The SDE* (4.10) *admits a weak solution such that*

(4.11) $E\int_0^T |U_t|^2 dt < \infty,$

and $L(U_.)$ *is uniquely determined by* Q, μ_0, *and* $\eta_.$.

To show that $B[\eta_.,\mu_0]$ is well-posed we need to prove uniqueness of the solution to (4.10) in the class of cadlag processes. The next result, Lemma 3.3(iv) of Funaki [7], does it.

LEMMA 4.4. *Let* $\eta_. \in B_2$, $\mu_0 \in P_2(S)$, *and* $P \in B[\eta_.,\mu_0]$; *let* $\mu_t = L(U_t)$. *Then* $t \longrightarrow \mu_t$ *belongs to* $C_{P_2(S)}[0,\infty)$.

Hence (4.10) implies (4.11), and the preceding three results yield

THEOREM 4.5. *The* $B[\eta_.,\mu_0]$-*martingale problem is well-posed.*

REMARK 4.2. A simple time substitution shows that, if V_t, $t \geq s$, is a process such that $V_s = v$ and

$$g(V_t) - g(V_s) - \int_s^t \langle \eta_r, Kg(V_r,\cdot)\rangle dr, \qquad t \geq s,$$

is a martingale for each $g \in C_c^2(S)$, then $L(V_.)$ is uniquely determined.

THE BOLTZMANN EQUATION

The following result, proven in App. B, is based on the time-inhomogeneous version of [EK], ch. 4, Proposition 9.19, given in App. B.

THEOREM 4.6. *For* $\eta_. \in B_2$ *and* $\mu_0 \in P_2(S)$, *the equation*

(4.12) $$\langle \mu_t, f\rangle = \langle \mu_0, f\rangle + \int_0^t \langle \mu_s \otimes \eta_s, Kf\rangle ds, \qquad f \in C_c^2(S),$$

admits a unique solution μ_t *in* B_2, *viz.,* $\mu_t = P \circ U_t^{-1}$, P *being the unique solution of* $B[\eta_.,\mu_0]$.

THEOREM 4.7. *For* $\mu_0 \in P_2(S)$, *the* $B[\mu_0]$-*martingale problem is well-posed.*

PROOF. By definition, $P \in B[\mu_0]$ iff $P \in B[\eta_\cdot, \mu_0]$ and $P \circ U_t^{-1} = \eta_t$, $t \geq 0$. Thus, by Theorem 4.2 and Lemma 4.4, $P \in B[\mu_0]$ iff P is a weak solution to (4.10) with $L(U_t) = L(Y_t)$ and (4.11) holds for all $T < \infty$. Existence and uniqueness of such a solution to (4.10) follows from Theorem 4 of Tanaka [22], modified to bring in ϕ in place of σ in (3.9) of [22].

Putting together the above results, we have the following important result.

THEOREM 4.8. *For* $\nu_0 \in P_2(S)$, *there is a unique solution* $\{\nu_t\}$ *in* B_2 *to the Boltzmann equation*

$$(4.13) \qquad \langle \nu_t, f \rangle = \langle \nu_0, f \rangle + \int_0^t \langle \nu_s \otimes \nu_s, Kf \rangle ds, \qquad f \in C_c^2(S),$$

namely, $\nu_t = \tilde{P} \circ U_t^{-1}$, *where* $\tilde{P}$ *is the unique solution of* $B[\nu_0]$.

PROOF. Let $\tilde{P} \in B[\nu_0]$ and $\mu_t = \tilde{P} \circ U_t^{-1}$; then μ_t satisfies (4.12) with $\eta_s = L(Y_s) = \mu_s$, i.e., μ_t satisfies (4.13).

For uniqueness, let $\nu_\cdot \in B_2$ be a solution to (4.13), let P^* be the unique solution to $B[\nu_\cdot, \nu_0]$, and let $\mu_t = P^* \circ U_t^{-1}$. Then $\mu_\cdot \in B_2$ and $\mu_\cdot$ solves (4.12) with $\eta_t = \nu_t$. But ν_t also solves (4.12) with $\eta_t = \nu_t$, whence $\mu_t = \nu_t$ by Theorem 4.6. Thus $\nu_t = \mu_t = P^* \circ U_t^{-1}$, hence $P^* \in B[\nu_0]$, i.e., $P^* = \tilde{P}$, and the proof is complete.

COROLLARY 4.9. *The function* $t \longrightarrow \nu_t$ *belongs to* $C_{P_2(S)}[0,\infty)$.

This is immediate by Lemma 4.4.

For $t \geq 0$, define $T_t: P_2(S) \to P_2(S)$ by $T_t\nu = \nu_t$, where ν_t is the unique solution in B_2 to the B.E. (4.13) with $\nu_0 = \nu$. It follows that T_t is a (nonlinear) semigroup, $T_tT_s = T_{t+s}$, called the

Boltzmann semigroup.

THE L-LOCAL MARTINGALE PROBLEM

Recall the definition (3.22) of L and the bound (3.23) on LH_{fg}. Since LH_{fg} may not be bounded, we consider the L-local martingale problem, as opposed to martingale problem.

A cadlag process $\beta_t = (U_t, \xi_t)$ with values in $E = S \times P_2(S)$ is *a solution to the* (L,μ)*-local martingale problem* if $L(\beta_0) = \mu$ $(\mu \in P(E))$ and, for $H = H_{fg} \in E_0$ (E_0 is defined after (3.16)),

$$(4.14) \qquad Y_{fg}(t) := H(\beta_t) - H(\beta_0) - \int_0^t LH(\beta_s)ds$$

is a local martingale; we write $\beta_\cdot \in (L,\mu)_{loc}$ or $L(\beta_\cdot) \in (L,\mu)_{loc}$ for this.

In proving that the L-local martingale problem is well-posed, we begin with degenerate initials.

THEOREM 4.10. *Let* $z = (u_0, \nu_0) \in E$.

a) *There exists a unique solution to the* (L,δ_z)*-local martingale problem.*

b) *Let* $\beta_\cdot = (U_\cdot, \xi_\cdot) \in (L,\delta_z)_{loc}$; *then*

$$(4.15) \qquad \xi_t = T_t(\nu_0) \quad \textit{for all} \quad t \geq 0, \quad \text{a.s.}$$

$$(4.16) \qquad U_\cdot \in B[\nu_\cdot, \delta_{u_0}], \quad \textit{where} \quad \nu_t = T_t(\nu_0).$$

PROOF. By Remark 4.3 below, we can choose $z_n = (u_n, \nu^n) \in E_n$ such that $z_n \longrightarrow z$, thus the existence in (a) follows from Theorem 3.4.

To get (4.15) notice that, for $g \in C_c^2(S)$, $L(H_{1g}^2) = 2H_{1g}LH_{1g}$ by (3.22), whence both $Y_{1g}(t)$ and $Y_{1g}^2(t)$ are local martingales vanishing at $t = 0$. As in the proof of Lemma 2.4, $Y_{1g}(t) \equiv 0$ a.s. This says

(4.17) $\exp\langle g,\xi_t\rangle = \exp\langle g,\xi_0\rangle + \int_0^t \exp\langle g,\xi_s\rangle\langle\xi_s \otimes \xi_s, Kg\rangle ds,\ t \geq 0,$ a.s.

A separability argument shows that (4.17) holds for all $g \in C_c^2(S)$ and $t \geq 0$, a.s., and then a simple differentiation argument (replace g by rg and compute $d\langle rg,\xi_t\rangle/dr$ at $r = 0$) proves that ξ_t satisfies the B.E. (4.13) a.s. By Theorem 4.7, $\xi_t = \nu_t := T_t(\nu_0)$ for all $t \geq 0$, a.s., i.e., (4.15) holds.

Taking $H = H_{f,0}$ in (4.14) shows that $U_\cdot$ solves the $B[\nu_\cdot,\delta_{u_0}]$-local martingale problem, hence the $B[\nu_\cdot,\delta_{u_0}]$-martingale problem, as noted after (4.6), so (4.16) is proven.

The uniqueness in (a) now follows because $L(\beta_\cdot)$ is completely determined by (4.15)-(4.16).

REMARK 4.3. Given $z = (u,\nu) \in E$, there exist $z_n \in E_n$ such that $z_n \longrightarrow z$ in E: let $X_1 = u$ and let $X_2, X_3, \ldots$ be i.i.d. rvs with law ν on S. Then, with $z_n := (u, n^{-1}\sum_1^n \delta_{X_i})$, we have $z_n \longrightarrow z$ a.s. by the law of large numbers. Moreover, for any $\mu \in P(E)$, there exist $\mu_n \in P(S^n)$ such that $\mu_n \circ R_n^{-1} \longrightarrow \mu$ in $P(E)$.

Let $\bar{P}_z$ be the unique solution on $\Omega = D_E[0,\infty)$ of the (L,δ_z)-local martingale problem. By (4.3), (4.8), and (4.15), $\bar{P}_z$ actually solves the (L,δ_z)-*martingale* problem (cf. also Theorem 3.4). As in Remark 2.1, $z \longrightarrow \bar{P}_z$ is measurable.

THEOREM 4.11. *For* $\mu \in P(E)$, *the* (L,μ)-*local martingale problem is well-posed, with unique solution (on* Ω*)*

$$\bar{P}_\mu(B) = \int_E \bar{P}_z(B)\mu(dz).$$

PROOF. Let $Z_t = (U_t,\xi_t)$ be the coordinate function on Ω and define $Y_{fg}(t)$ by (4.14) but with $Z_\cdot$ instead of $\beta_\cdot$. Then, for each z,

$b^2(\xi_{t\wedge\tau_N}) \leq N$ $\bar{P}_z$ - a.s. because of (4.15) and Corollary 4.9. Using $|H_{fg}(u,\nu)| \leq \|f\|e^{\|g\|}$ and (4.3) we find

$$(4.18) \qquad |LH_{fg}(u,\nu)| \leq C_{fg}(1 + b^2(\nu)),$$

where $C_{fg} = e^{\|g\|}(C_f + \|f\|C_g)$, whence

$$(4.19) \qquad |Y_{fg}(t\wedge\tau_N)| \leq 2\|f\|e^{\|g\|} + tC_{fg}(1 + N), \quad \bar{P}_z \text{ - a.s.}$$

This implies that $Y_{fg}(\cdot\wedge\tau_N)$ is a $\bar{P}_\mu$-martingale, so that $Y_{fg}(\cdot)$ is a $\bar{P}_\mu$-local martingale, i.e., $\bar{P}_\mu \in (L,\mu)_{loc}$.

Next, if $P \in (L,\mu)_{loc}$, it follows, as in the proof of 4.10, that $\xi_t(\omega) = T_t(\xi_0(\omega))$, $t \geq 0$, P - a.s. Thus, by 4.9, $t \longrightarrow \xi_t(\omega)$ is continuous P - a.s., and (4.19) holds, so $Y_{fg}(\cdot\wedge\tau_N)$ is a martingale.

Let $P_{\omega'}$ be a regular version of $P(\cdot|\sigma(Z_0))$; then, for P - almost all ω', $Y_{fg}(\cdot\wedge\tau_N)$ is a $P_{\omega'}$-martingale for all $H_{fg} \in E_0$, $N \geq 1$. For such ω', $P_{\omega'} = \bar{P}_{Z_0(\omega')}$, whence

$$P(B) = \int \bar{P}_{Z_0(\omega')}(B)P(d\omega') = \bar{P}_\mu(B) .$$

COROLLARY 4.12) a) *Let* $\mu_0 \in P(S)$, $\nu_0 \in P_2(S)$, *and* $\mu = \mu_0 \times \delta_{\nu_0} \in P(E)$; *then, under* $\bar{P}_\mu$, $U_\cdot \in B[\nu_\cdot,\mu_0]$, *where* $\nu_t = T_t(\nu_0)$.

b) *If* $\mu_0 = \nu_0$ *in* (a), *then* $U_\cdot \in B[\nu_0]$.

The proof of (a) is similar to that of (4.16). For (b), let $\nu_n^{\otimes n}$ be the n-fold product of ν_0 on S^n and let β_0^n have law $L(\beta_0^n) = \nu_0^{\otimes n} \circ R_n^{-1}$. Then $\beta_\cdot^n \xrightarrow{D} \beta_\cdot$, where $\beta_t = (U_t,\nu_t)$, $\nu_t = T_t(\nu_0)$. Thus $\nu_t^n = n^{-1}\sum_1^n \delta_{V_i^n(t)} \xrightarrow{D} \nu_t$ in $P_2(S)$, and we may assume that $\nu_t^n \xrightarrow{\rho_2} \nu_t$ a.s. This implies $\langle\nu_t^n,g\rangle \to \langle\nu_t,g\rangle$ for $g \in C_b(S)$ a.s. But (cf. the proof of Theorem 3.1) $V^n(t) = (V_1^n(t),\ldots,V_n^n(t))$ is exchangeable, thus

$E\langle\nu_t^n,g\rangle = Eg(U_t^n)$. It follows that $\nu_t = \mathcal{L}(U_t)$, so $U_\cdot \in B[\nu_0]$.

A TIME-INHOMOGENEOUS MARKOV PROCESS

Let $F_k = S \times \mathcal{P}_k$, $k = 0,1,\ldots$, where

$$\mathcal{P}_k = \{\mu \in \mathcal{P}_2(S) : k \leq b^2(\mu) < k+1\}\ .$$

Define L^k to be L acting on the domain $\mathcal{E}_k$ of function $H_{fg} \in \mathcal{E}_0$ restricted to F_k.

The arguments above show that the L^k-martingale problem is well-posed. Since L^k is a bounded operator, the solution is a (time-homogeneous) Markov process with state space F_k, by [EK], ch. 4, 4.2. Gluing these processes together in an obvious way gives a Markov process with state space E, namely $\beta_\cdot$, the solution of the L-local martingale problem.

Let $P_{u,\nu}$, $u \in S$, $\nu \in \mathcal{P}_2(S)$, denote the solution of the $(L,\delta_{(u,\nu)})$-martingale problem, and put

(4.20) $$e_\nu(s,u,A) = P_{u,\nu}(U_s \in A), \quad s \geq 0, \ A \in \mathcal{B}(S).$$

It is easy to check that (4.20) is the transition function associated with the time-inhomogeneous Markov process U_t in the sense of Tanaka [20]; see also Mc Kean [15].

CLASSICAL PROPERTIES OF SOLUTIONS OF THE BOLTZMANN EQUATION

Here we derive the preservation of second moments, convergence to equilibrium, and propagation of chaos as consequences of the n-particle approximation. Let $\nu_t = T_t(\nu_0)$, as in Theorem 4.8.

THEOREM 4.13. *Let* $\nu_0 \in \mathcal{P}_2(S)$; *then* $b^2(\nu_t) = b^2(\nu_0)$, $t \geq 0$.

As in the proof of Corollary 4.12(b), we have $\nu_t^n \xrightarrow{\rho_2} \nu_t$ a.s., thus $b^2(\nu_t^n) \to b^2(\nu_t)$ a.s. . By Lemma 2.4, $b^2(\nu_t^n) = b^2(\nu_0^n) \to b^2(\nu_0)$ a.s., and the result follows.

Using estimates like (2.27), it can be shown that, for some constants C_1, C_2,

$$\int |v|^{2k} \nu_t(dv) \leq C_1 e^{C_2 t} \int |v|^{2k} \nu_0(dv).$$

Let $Q^\varepsilon(B) = Q(B \cap [\varepsilon,\pi])$, as in §2; thus $Q^\varepsilon((0,\pi]) < \infty$ and we are in the so-called cut-off case. The Boltzmann semigroup given by Theorem 4.8 is denoted by T_t^ε.

THEOREM 4.14. *Let* $\nu_0 \in P_2(S)$, $0 < T < \infty$; *then*

$$\rho_2(T_t^\varepsilon \nu_0, T_t \nu_0) \to 0 \quad \text{as } \varepsilon \downarrow 0 \tag{4.21}$$

uniformly for $0 \leq t \leq T$.

PROOF. Let $\mu^n = \nu^{\otimes n}$ (cf. 4.12(b)). Using Theorem 2.7, get $W^n \in (K_n, \mu^n)$ and $W^{n,\varepsilon} \in (K_n^\varepsilon, \mu^n)$ such that (2.34) holds. Let $\nu_t^n = n^{-1} \sum_1^n \delta_{W_i^n(t)}$, $\nu_t^{n,\varepsilon} = n^{-1} \sum_1^n \delta_{W_i^{n,\varepsilon}(t)}$. Then $\mu^n \circ R_n^{-1} \to \nu_0 \times \delta_{\nu_0}$ implies $\nu_\cdot^n \to \nu_\cdot$ and $\nu_\cdot^{n,\varepsilon} \to \nu_\cdot^\varepsilon$ in $D_{P_2(S)}[0,\infty)$ in probability, and $\nu_t = T_t(\nu_0)$, $\nu_t^\varepsilon = T_t^\varepsilon(\nu_0)$. Since ν_t and ν_t^ε are in $C_{P_2(S)}[0,\infty)$, we get

$$\sup_{t \leq T} \rho_2(\nu_t^n, \nu_t) \to 0 \quad \textit{in prob.} \tag{4.22}$$

$$\sup_{t \leq T} \rho_2(\nu_t^{n,\varepsilon}, \nu_t) \to 0 \quad \textit{in prob.} \tag{4.23}$$

On the other hand,

$$\rho_2^2(\nu_t^n, \nu_t^{n,\varepsilon}) \leq \frac{1}{n} |W^n(t) - W^{n,\varepsilon}(t)|^2 ,$$

hence, by (2.34),

(4.24) $$E \sup_{t\leq T} \rho_2^2(\nu_t^n, \nu_t^{n,\varepsilon}) \leq C_T \frac{1}{n}\int |\underline{v}|^2 \mu^n(d\underline{v}) \int_0^\varepsilon \theta^2 Q(d\theta)$$

$$\leq C_T \int |u|^2 \nu_0(du) \int_0^\varepsilon \theta^2 Q(d\theta).$$

Letting $n \to \infty$ in (4.24), and using (4.22-4.23) and Fatou's lemma, we get

$$\sup_{t\leq T} \rho_2^2(T_t(\nu_0), T_t^\varepsilon(\nu_0)) \leq C_T b^2(\nu_0) \int_0^\varepsilon \theta^2 Q(d\theta),$$

which gives the required result.

As a consequence we have

THEOREM 4.15. *Let* $\nu_0, \nu_0' \in P_2(S)$. *Then*

(4.25) $$\rho_2(T_t(\nu_0), T_t(\nu_0')) \leq \rho_2(\nu_0, \nu_0'), \qquad t \geq 0,$$

(4.26) $$\rho_2(T_t(\nu_0), G(\nu_0)) \to 0 \qquad (t \to \infty),$$

where $G(\nu_0)$ *is the Gaussian distribution on* S *with density*

$$(2\pi\sigma^2)^{-3/2} \exp(-|v - m|^2/2\sigma^2),$$

where $m = \int v\nu_0(dv)$, $3\sigma^2 = \int |v - m|^2 \nu_0(dv)$.

PROOF. The proof in §§7,9 of Tanaka [20], for the case $\int \theta dQ < \infty$, works here in view of (4.21) (cf. [20],Lemma 7.2(ii)).

PROPAGATION OF CHAOS

Ignoring the velocity components of $\beta_\cdot^n$ and $\beta_\cdot$ in our earlier analysis we get the following result. Let $V^n \in (K_n)$, $\nu_t^n = n^{-1} \sum_1^n \delta_{V_i^n(t)}$, and suppose $\nu_0^n \xrightarrow{\rho_2} \nu_0$ in probability. Then (4.22) holds with $\nu_t = T_t(\nu_0)$. When the law of $V^n(0)$ is symmetric on S^n, this implies the classical propagation of chaos, (a stronger version is in Theorem 4.16 below), but the present result does not require symmetry.

Next, we consider convergence of the n-particle process V^n to an infinite-particle process. Let $V^n \in (K_n)$ and

$$X_t^n := (V_1^n(t),\dots,V_n^n(t),0,0,\dots;\nu_t^n),$$

so X^n has state space $E^* := S^\infty \times P_2(S)$. Suppose

$$(4.27) \qquad X_0^n \xrightarrow{D} \Lambda \in P(E^*) .$$

Then it can be shown that each component sequence $\{V_i^n(\cdot)\}$ (i fixed) is relatively compact, hence $\{X^n\}$ is likewise. Further, any subsequential limit X is a solution to the L^∞-local martingale problem, where the domain E^∞ of L^∞ consists of functions

$$(4.28) \qquad h(\underline{u},\nu) := H_{f_1,\dots,f_k,g}(\underline{u},\nu) = e^{<\nu,g>} \prod_1^k f_i(u_i),$$

with $\underline{u} = (u_1,u_2,\dots) \in S^\infty$, $\nu \in P_2(S)$, $f_i \in C_c^2(S)$ or $f_i = 1$, $g \in C_c^2(S)$, and

$$(4.29) \qquad L^\infty h(\underline{u},\nu) = \sum_{j=1}^k LH_{f_j g}(u_j,\nu) \prod_{i\neq j} f_i(u_i).$$

Thus, if $X_t = (U_1(t),U_2(t),\dots;\nu_t)$ is a solution to the L^∞-local martingale problem, then $(U_i(\cdot),\nu_\cdot) \in (L)_{loc}$ for each i. If ν_0 is nonrandom, then $\nu_t = T_t(\nu_0)$, and a time-inhomogeneous version of [EK], ch. 4, 10.1, shows that, if $U_1(0),U_2(0),\dots$ are independent, then the processes $U_1(\cdot),U_2(\cdot),\dots$ are likewise. Writing $\bar{P}_{u,\nu}$ for the solution to $(L,\delta_{(u,\nu)})_{loc}$, we get the following. Suppose $X_\cdot \in (L^\infty,\delta_{(\underline{u},\nu)})_{loc}$; then $\nu_t = T_t(\nu)$, $t \geq 0$, a.s., and, for $A_i \in B(D_S[0,\infty))$,

$$(4.30) \qquad P(U_i(\cdot) \in A_i\colon 1 \leq i \leq k) = \prod_{i=1}^k \bar{P}_{u_i,\nu}(A_i).$$

Thus $(L^\infty, \delta_{(\underline{u},\nu)})_{loc}$ is well-posed. As in Remark 2.1, the L^∞-local martingale problem is well-posed.

It follows that, if (4.27) holds, then $X^n \xrightarrow{D} X_\cdot \in (L^\infty, \Lambda)_{loc}$, and the law of $X_\cdot$ is determined, for A_i as in (4.30), $B \in \mathcal{B}(D_{P_2(S)}[0,\infty))$, by

$$(4.31) \quad P(U_i(\cdot) \in A_i,\ 1 \leq i \leq k,\ \nu_\cdot \in B)$$

$$= \int_{E^*} I_{\{T_\cdot(\nu) \in B\}} \prod_{i=1}^{k} \bar{P}_{u_i,\nu}(A_i)\Lambda(d\underline{u}d\nu).$$

Now suppose that (4.27) holds and that $L(V^n(0))$ is symmetric.

It follows from de Finetti's theorem that Λ admits a representation

$$(4.32) \quad \Lambda(A \times B) = \int_B \gamma^\infty(A)\Gamma(d\gamma),\quad A \in \mathcal{B}(S^\infty),\quad B \in \mathcal{B}(P_2(S)),$$

where Γ is a probability measure on $P_2(S)$ and γ^∞ is the infinite product measure $\gamma \otimes \gamma \otimes \ldots,\ \gamma \in P_2(S)$.

Thus, for $f_i \in C_b(S)$,

$$(4.33) \quad E\left[\prod_{i=1}^{k} f_i(V_i^n(0))\right] \to \int \prod_{i=1}^{k} \langle \gamma, f_i \rangle \Gamma(d\gamma),$$

and, for $H_1,\ldots,H_k \in C_b(D_S[0,\infty))$,

$$(4.34) \quad E\left[\prod_{i=1}^{k} H_i(V_i^n(\cdot))\right] \to \int \prod_{i=1}^{k} \langle P_\gamma^*, H_i \rangle \Gamma(d\gamma),$$

where P_γ^* is the unique solution of the Boltzmann martingale problem $B[\gamma]$. When $\Gamma = \delta_\gamma$, (4.34) implies that $V_1^n(\cdot),\ldots,V_m^n(\cdot)$ are asymptotically (as $n \to \infty$, m fixed) i.i.d. with law P_γ^*. This is a form of "functional propagation of chaos" (see also Sznitman [19]). Here is the formal statement of this result.

THEOREM 4.16. *Let* μ_n *be symmetric probability laws on* S^n *which are "*ν*-chaotic",* $\nu \in P_2(S)$, i.e., *for* $f_i \in C_b(S)$, $1 \leq i \leq k$,

$$(4.35) \qquad \langle \mu_n, f_1 \otimes \dots \otimes f_k \otimes 1 \otimes 1 \otimes \dots \rangle \to \prod_{i=1}^{k} \langle f_i, \nu \rangle ,$$

and suppose

$$(4.36) \qquad \frac{1}{n}\int_{S^n} |\underline{v}|^2 \mu_n(d\underline{v}) \to \int_S |u|^2 \nu(du).$$

Let $\tilde{P}_n \in (K_n, \mu_n)$; *then the* $\tilde{P}_n$ *are* P_ν*-chaotic, where* $P_\nu \in B[\nu]$, i.e.,

$$(4.37) \qquad \langle \tilde{P}_n, H_1 \otimes \dots \otimes H_k \otimes 1 \otimes \dots \rangle \to \prod_{i=1}^{k} \langle P_\nu, H_i \rangle,$$

for $H_i \in C_b(D_S[0,\infty))$, $1 \leq i \leq k$.

APPENDIX A

All references here are to [J].

Let $(E,\mathcal{E})$ be a Lusin space and λ a measure on $\mathcal{E}$. Let $a(t,u,z)$ and $b(t,u)$ be measurable maps from $[0,\infty) \times \mathbb{R}^m \times E$ $([0,\infty) \times \mathbb{R}^m)$ into $\mathbb{R}^m$. Suppose that

(A.1) $\quad b(t,u)$ *is bounded on compacts*

(A.2) $\quad \int |a(t,u,z)|^2 \lambda(dz)$ *is bounded on compacts.*

Let $A(t,u,dy)$ be the (not necessarily finite) kernel on $[0,\infty) \times \mathbb{R}^m \times \mathcal{B}(\mathbb{R}^m)$ given by

$$\int_{\mathbb{R}^m} g(y)A(t,u,dy) = \int_E g(a(t,u,z))I_{\{a(t,u,z)\neq 0\}}\lambda(dz).$$

Condition (A.2) implies that $\int |y|^2 A(t,u,dy)$ is bounded on compacts.

Finally, define operators A_t taking $C_b^2(\mathbf{R}^m)$ into Borel functions on $\mathbf{R}^m$ by

(A.3) $$A_t g(u) = \nabla g(u) \cdot b(t,u) + \int_{\mathbf{R}^m} (g(u+y) - g(u) - \nabla g(u) \cdot y) A(t,u,dy).$$

Since the integrand is dominated by $|||g|||\ |y|^2$ (see (2.11) for $|||\cdot|||$), $A_t g(u)$ is well-defined. Moreover,

(A.4) $$|A_t g(u)| \leq |||g|||\ \{|b(t,u)| + \int |y|^2 A(t,u,dy)\} .$$

Let $f'(y) := yI_{\{|y|\leq 1\}}$, $f''(y) := y - f'(y)$. Then

$$A_t g(u) = \nabla g(u) \cdot \beta(t,u) + \int_{\mathbf{R}^m} (g(u+y) - g(u) - f'(y) \cdot \nabla g(u)) A(t,u,dy),$$

where $\beta(t,u) := b(t,u) - \int_{\mathbf{R}^m} f''(y) A(t,u,dy)$. In view of A.2), $\beta(t,u)$ is well-defined.

Let $\tilde{\Omega} = D_{\mathbf{R}^m}[0,\infty)$ with canonical process $\tilde{X}_t$, canonical (right continuous) filtration $\tilde{F}_t$, and $\tilde{F} = \tilde{F}_\infty$.

THEOREM A.1. *For a probability measure* $\tilde{P}$ *on* $(\tilde{\Omega},\tilde{F})$, *the following three statements are equivalent:*

(i) $Mg(t) := g(\tilde{X}_t) - g(\tilde{X}_0) - \int_0^t A_s G(\tilde{X}_s) ds$ *is a local martingale for each* $g \in C_b^2(\mathbf{R}^m)$.

(ii) $\tilde{X}$ *is a semimartingale with local characteristics* (B,C,ν) *given by*

$$B_t = \int_0^t \beta(s,\tilde{X}_s) ds, \quad C_t \equiv 0, \quad \nu(\tilde{\omega},dtdy) = A(t,\tilde{X}_t(\tilde{\omega}),dy)dt .$$

(iii) $\tilde{P}$ *is a weak solution of the SDE*

(A.5) $$dX_t = b(t,X_t)dt + a(t,X_{t-},z)\tilde{N}(dtdz),$$

where N *is a Poisson measure on* $\mathbf{R}_+ \times E$ *with intensity* $\lambda_N(dtdz) = dt\lambda(dz)$ *and* $\tilde{N} = N - \lambda_N$ *is its compensated version.*

Further, if $A_t g(u)$ *is bounded in* $(t,u) \in [0,T] \times \mathbf{R}^m$, *then*

(i) *is equivalent to*

(i') $Mg(t)$ *is a martingale for* $g \in C_c^2(\mathbf{R}^m)$.

PROOF. (i) is equivalent to (ii) by (13.55). From (14.80), (ii) holds iff $\tilde{P}$ is a weak solution of

$$dX_t = \beta(t,X_t)dt + f'(a(t,X_{t-},z))\tilde{N}(dtdz) + f''(a(t,X_{t-},z))N(dtdz), \tag{A.6}$$

with N as in (iii). Adding and subtracting

$$\int_0^s\int f''(a(t,X_t,z))dt\lambda(dz) = \int_0^s (b(t,X_t) - \beta(t,X_t))dt$$

in the integrated version of (A.6) one gets that $\tilde{P}$ is a weak solution to (A.5) iff it is a weak solution to (A.6).

The last assertion follows by approximating $g \in C_b^2(S)$ by functions in $C_c^2(S)$.

APPENDIX B

All references in this appendix are to [EK].

Let E be a locally compact, separable metric space. The space of bounded measurable functions on E is $B(E)$. Notions related to bounded pointwise (= bp) convergence are in [EK], App. 3. We write $C_0(E)$ for the continuous functions vanishing at infinity.

For $t \geq 0$, let $A_t : \mathcal{D} \to B(E)$ be linear operators on $\mathcal{D} \subset C_0(E)$. Suppose

a) $\mathcal{D}$ is an algebra uniformly dense in $C_0(E)$

b) for each $t \geq 0$, A_t satisfies the positive maximum principle

c) $(t,x) \longrightarrow A_t f(x)$ is measurable for $f \in \mathcal{D}$

d) there is a countable subset $\mathcal{D}_0$ of $\mathcal{D}$ such that, for each t,

(B.1) $\{(f,A_tf): f \in \mathcal{D}\} \subset$ bp-closure $\{(f,A_tf): f \in \mathcal{D}_0\}$.

We say the (A_t)-*martingale problem is well-posed* if, for each $(s,x) \in E' := [0,\infty) \times E$, there is a unique probability measure $Q_{s,x}$ on $(D_E[0,\infty), \sigma(X_t: t \geq s))$, X_t being the coordinate process, such that

$$f(X_t) - f(X_s) - \int_s^t A_r f(x)dr, \quad t \geq s, \quad f \in \mathcal{D}$$

is a $Q_{s,x}$-martingale.

THEOREM B.1. *Suppose that the* (A_t)-*martingale problem is well-posed. Then, given* $\mu \in P(E)$, *the equation*

$$\text{(B.2)} \quad \langle\mu_t,f\rangle = \langle\mu,f\rangle + \int_0^t \langle\mu_r,A_rf\rangle dr, \quad f \in \mathcal{D}$$

has a unique solution $\{\mu_t\}$ *among Markov kernels on* $[0,\infty) \times B(E)$.

PROOF. Let $\mathcal{D}'$ be the algebra of functions of the form

$$\text{(B.3)} \quad g(t,x) = \sum_{i=1}^{k} h_i(t)f_i(x),$$

where $h_1,\ldots,h_k \in C_c^1([0,\infty))$ and $f_1,\ldots,f_k \in \mathcal{D}$. Note that $\mathcal{D}'$ is dense in $C_0(E')$.

Define $A' := \partial/\partial t + A_t$ on $\mathcal{D}'$, i.e., for $g \in \mathcal{D}'$,

$$\text{(B.4)} \quad A'g(t,x) = \sum_{i=1}^{k}\left\{\left[\frac{\partial}{\partial t} h_i(t)\right]f_i(x) + h_i(t)A_tf_i(x)\right\}.$$

Suppose $g(t,x)$ attains its maximum at (t_0,x_0). Since $g(t_0,x)$ is in $\mathcal{D}$ and has a maximum at x_0, $A_{t_0}g(t_0,x_0) \leq 0$. On the other hand, $g(t,x_0)$ is in $C_c^1([0,\infty))$ and has a maximum at t_0, so $\frac{\partial}{\partial t}g(t_0,x_0) \leq 0$. Hence $A'g(t_0,x_0) \leq 0$.

Thus A' satisfies the positive maximum principle.

Arguing as in [EK], ch. 4, Theorem 7.1, we find that the $(A',\delta_{(s,x)})$-martingale problem is well-posed for each $(s,x) \in E'$.

Further, if $\mathcal{D}'_0$ is the countable collection of functions g in (B.3) with $f_i \in \mathcal{D}_0$ and h_i belonging to a countable dense subset of $C^1_c([0,\infty))$ (in the norm $\|h\|_1 := \|h\| + \|\frac{\partial}{\partial t} h\|$), then, by (B.1),

(B.5) $$\{(g,A'g): g \in \mathcal{D}'\} \subset \text{bp-closure } \{(g,A'g): g \in \mathcal{D}'_0\} .$$

Thus, as in Remark 2.1, A' is separable and the A'-martingale problem is well-posed.

Let μ_t be a solution of (B.2) and define $\lambda_t = \delta_t \otimes \mu_t$. For $f \in \mathcal{D}$, $h \in C^1_c([0,\infty))$, observe that

$$<\lambda_t,fh> = h(t)<\mu_t,f>,$$

which is an absolutely continuous function of t. Thus

$$\begin{aligned} <\lambda_t,fh> &= <\lambda_0,fh> + \int_0^t \frac{\partial}{\partial s} [h(s)<\mu_s,f>]ds \\ &= <\lambda_0,fh> + \int_0^t <\lambda_s,A'(fh)>ds. \end{aligned}$$

By linearity of A' we get

$$<\lambda_t,g> = <\lambda_0,g> + \int_0^t <\lambda_s,A'g>ds, \quad g \in \mathcal{D} ;$$

thus, by ch. 4, proposition 9.19, λ_t is uniquely determined by $\lambda_0 = \delta_0 \times \mu$, hence μ_t is uniquely determined by $\mu_0 = \mu$.

PROOF OF THEOREM 4.6. Take $E = S$ $(= \mathbf{R}^3)$, $\mathcal{D} = C^2_c(S)$, and

$$A_t f(u) = \int Kf(u,v)\eta_t(dv)$$

in Theorem B.1. By (4.3), $A_t f(u)$ is bounded. It is easy to check the positive maximum principle. The estimate (3.26) implies (B.1) if we take $\mathcal{D}_0$ to be a countable dense subset of $C_c^2(S)$ in $|||\cdot|||$-norm. Well-posedness of the (A_t)-martingale problem is noted in Remark 4.2. Thus (4.12) has a unique solution among Markov kernels. Since $\mu_t = P \circ U_t^{-1}$ is a solution in $\mathcal{B}_2$, it follows that (4.12) has a unique solution in $\mathcal{B}_2$.

REFERENCES

1. R. BASS, Uniqueness in law for pure jump Markov processes, to appear in PTRF.
2. C. CERCIGNANI, *The Boltzmann Equation and its Applications.* Springer 1988.
3. P. ECHEVERRIA, A criterion for invariant measures of Markov processes, *ZW* 61, 1982, 1-16.
4. S. ETHIER and T. KURTZ, *Markov Processes: Characterization and Convergence.* Wiley 1986.
5. T. FUNAKI, The diffusion approximation of the Boltzmann equation of Maxwellian molecules. *Publ. RIMS, Kyoto Univ.* 19, 1983, 841-886.
6. T. FUNAKI, A certain class of diffusion processes associated with nonlinear parabolic equations. *ZW* 67, 1984, 331-348.
7. T. FUNAKI, The diffusion approximation of the spatially homogeneous Boltzmann equation. *Duke Math. J.* 52, 1985, 1-23.
8. P. GÉRARD, Solutions globale du problème de Cauchy pour l'équation de Boltzmann. *Sem. Bourbaki* 40, 1987-88, no. 699.
9. F. A. GRÜNBAUM, Propagation of chaos for the Boltzmann equation. *Arch. Ratl. Mech. Anal.* 42, 1971, 323-345.
10. J. JACOD, *Calcul Stochastiques et Problèmes de Martingales. Springer Lect. Notes Math.* no. 714, 1979.
11. M. KAC, Foundations of kinetic theory. *Proc. Third Berk. Symp. on Math. Stat. Prob.* 3, 1956, 171-197.

12. T. KOMATSU, On the pathwise uniqueness of solutions of one- dimensional stochastic differential equations of jump type. *Proc. Japan. Acad. Sci.* ser. A, 58, 1982, 353-356.

13. J.-P. LEPELTIER et B. MARCHAL, Problèmes de martingales et équations différentielles stochastiques associées à un opérateur intégro-différentiel. *Ann. Inst. H. Poincaré*, Nouv. Ser. B, 12, 1976, 43-103.

14. R. LIPTSER and A. SHIRYAEV, *Statistics of Stochastic Processes*, Springer 1977.

15. H. P. McKEAN, Jr., A Class of Markov processes associated with nonlinear parabolic equations. *Proc. Nat. Acad. Sci.* 56, 1966, 1907-1911.

16. K. OELSCHLÄGER, A martingale approach to the law of large numbers for weakly interacting stochastic processes. *Ann. Prob.* 12, 1984, 458-479.

17. D. STROOCK, Diffusion processes associated with Lévy generators. *ZW* 32, 1975, 109-244.

18. D. STROOCK and S.R.S. VARADHAN, *Multidimensional Diffusion Processes.* Springer 1979.

19. A.-S. SZNITMAN, Équations de type Boltzmann, spatiàlement homogènes. *ZW* 66, 1984, 559-592.

20. H. TANAKA, Probabilistic treatment of the Boltzmann equation of Maxwellian molecules. *ZW* 46, 1978, 67-105.

21. H. TANAKA, Some probabilistic problems in the spatially homogeneous Boltzmann equation, in *Theory and Application of Random Fields*, G. Kallianpur (ed.), Springer Lect. Notes Control Info. Sci. no. 49, 1983.

22. H. TANAKA, Stochastic differential equations corresponding to the spatially homogeneous Boltzmann equation of Maxwellian and non-cutoff type. *J. Fac. Sci. Univ. Tokyo* Sec. IA, 34, 1987, 351-369.

23. C. THOMPSON, *Mathematical Statistical Mechanics.* Princeton Univ. Press 1972.

24. C. TRUESDELL and R. MUNCASTER, *Fundamentals of Maxwell's Kinetic Theory of a Simple Monatomic Gas.* Academic Press 1980.

25. K. UCHIYAMA, Derivation of the Boltzmann equation from particle dynamics, *Hiroshima Math. J.* 18, 1988, 245-297.

J. HOROWITZ
Department of Mathematics
and Statistics
University of Massachusetts
Amherst, MA 01003
U.S.A.
e-mail: joeh@math.umass.edu

R. L. KARANDIKAR
Indian Statistical Institute
Delhi Center
7, S.J.S. Sansanwal Marg
New Delhi - 110016
INDIA

PROBABILISTIC METHODS IN DIFFERENTIAL GEOMETRY[1]

by

PEI HSU[2]

Let M be a Riemannian manifold with metric g. The metric g determines a second order elliptic operator Δ called the Laplace-Beltrami operator. Assume that locally the metric is given by the matrix $ds^2 = g_{ij}dx^i dx^j$, then the operator Δ has the form

$$\Delta = \frac{1}{\sqrt{\det g}}\frac{\partial}{\partial x^i}\left(\sqrt{\det g}g^{ij}\frac{\partial}{\partial x^j}\right).$$

It is well-known to probabilists that every second order nondegenerate elliptic operator gives rise to a diffusion process. The process associated with the Laplace–Beltrami operator (actually half of it, $\Delta/2$) is called the Riemannian Brownian motion X on M. We want to study the geometric properties of M by investigating the behavior of the Riemannian Brownian motion. For some geometric problems the probabilistic approach is nicer, clearer, and more motivated than analytic approaches, because we can "see" effects of geometric conditions (e.g., bounds of curvatures) through the behavior of Brownian paths. I am going to explain a few applications of this kind.

Locally we can get the Riemannian Brownian motion simply by solving a stochastic differential equation

$$dX_t^i = \sigma_j^i(X_t)\,dB_t^j - \frac{1}{2}g^{jk}(X_t)\Gamma_{jk}^i(X_s)\,ds.$$

[$\sigma = (\sigma_j^i)$ is any smooth square root of the metric matrix $g = (g_{ij})$.] This stochastic

[1]This work was supported in part by NSF grant DMS-86-01977.

[2]On leave from the University of Illinois at Chicago.

differential equation can only be used locally. To obtain the whole process one has to glue local Brownian motions from one coordinate patch to another. The details of this construction are tedious and are not really important to us. Another way to look at Brownian motion on a Riemannian manifold is to say that it is a measure on the path space which solves a martingale problem:

For any C^2 function f on M, the process

$$M_t^f = f(X_t) - \frac{1}{2}\int_0^t \Delta f(X_s)\, ds$$

is a continuous local martingale.

One has to note that this local martingale may not run for all time t. Its lifetime may be finite (a trivial case being a bounded domain in R^n). In fact to know when Brownian motion has finite lifetime is an interesting geometric problem. We will come back to this point later.

Sometimes we need to know the quadratic variation process of the martingale M^f. By comparing M^{f^2} with the expression obtained by using Itô's formula on $f(X_t)^2$, we find that $d\langle M^f, M^f\rangle_t = |\nabla f|^2(X_t)\, dt$.

A pretty definition of the Riemannian Brownian motion is that it is the projection of the horizontal Brownian motion on the the bundle of orthonormal frames. The advantage of the horizontal Brownian motion is that it can be globally defined without referring to local coordinates. Also this definition will allow you get an explicit expression for the martingale M^f in the formulation of the martingale problem. For the purpose of geometric applications, it is quite sufficient to stay on the manifold.

In many problems it is important to compare the Brownian motion on a Riemannian manifold with certain curvature conditions, to the Brownian motion on a model manifold, say, a radially symmetric manifold. For this purpose we need a comparison theorem for stochastic differential equations. In its simplest form, the comparison theorem says:

Suppose that X_t, Y_t are solutions of the two stochastic differential equations

$$dX_t = \sigma(X_t)\, dB_t + b(X_t, \omega)\, dt,$$

$$dY_t = \sigma(Y_t)\, dB_t + c(X_t, \omega)\, dt.$$

Suppose also that $b(\cdot, \omega) \le c(\cdot, \omega)$ and that $X_0 \le Y_0$. Then we have $X_t \le Y_t$ for all $t \ge 0$.

I have left out some technical assumptions in the above statement, see [IW, Chapter IV].

Many geometric problems can be solved by studying the rate of escape of the Brownian motion from a fixed point. In fact, in one way or another, all problems we are going to discuss here involve studying how fast (or how slowly) Brownian paths travel. This is where the Ricci curvature and the sectional curvature of the manifold come into the picture.

If we push Brownian motion near a fixed point O to the tangent plane at O by the exponential map $\exp_O^{-1}$ and fix a polar coordinate system on the tangent plane, then Brownian motion can be decomposed into two parts: the radial motion $R_t = r(X_t)$ [r is the distance function from O], and the angular motion $\Theta_t = \theta(X_t)$. Let us look at the radial Brownian motion R_t more closely. The Laplace-Beltrami operator in polar coordinates is

$$\Delta = \frac{\partial^2}{\partial r^2} + m(r, \theta)\frac{\partial}{\partial r} + \Delta_\Theta.$$

[Δ_Θ involves only derivatives in the Θ variables.] We can write down (locally) the stochastic differential equation for the radial process R_t:

$$dR_t = d\beta_t + \frac{m(R_t, \Theta_t)}{2} dt,$$

where β is a one-dimensional Brownian motion (because $|\nabla r| \equiv 1$). From the above equation we see that the rate of escape of R_t depends on the behavior of the function $m(r,\theta) = \Delta r$. This function is important enough that it is well studied in differential geometry. It turns out $m(r,\theta)$ can be bounded from below by the (radial) Ricci curvature and from above by the sectional curvature. To be more precise,

Suppose that

$$\kappa_1(r)^2 \geq -\inf\{\mathrm{Ric}(x) : r(x) \leq r\},$$

and

$$\kappa_2(r)^2 \leq -\inf\{\mathrm{Sect}(x) : r(x) \leq r\}.$$

Let $G_i(r), i = 1,2$ be defined by

$$G_i''(r) = \kappa_i(r)^2 G_i(r), \qquad G_i(0) = 0, \quad G_i'(0) = 1.$$

Then we have

$$(n-1)\frac{G_2'(r)}{G_2(r)} \leq m(r,\theta) \leq (n-1)\frac{G_1'(r)}{G_1(r)}.$$

These inequalities are consequences of a host of comparison theorems in differential geometry (see [GW]).

Let us now discuss a few problems which can be solved by what we have so far. First we consider the stochastic completeness of a complete Riemannian manifold. Let $p(t,x,y)$ be the minimal heat kernel. M is said to be stochastically complete if

$$\forall x \in M, \quad \forall t > 0, \qquad \int_M p(t,x,y)\,dy = 1.$$

Stochastic completeness is equivalent to the uniqueness of solutions of the initial value problem for the heat equation with $(L^\infty(M))$ initial data, see [IW, §V.3].

Let e be the lifetime of the Brownian motion. For a complete manifold, e is the explosion time of the radial process R, i.e., $e = \lim_{n\to\infty} e_n$, where $e_n = \inf\{t > 0 : r(X_t) \geq n\}$. We have

$$P_x\{e \geq t\} = \int_M p(t,x,y)\,dy,$$

M is stochastically completely if and only if Brownian motion has infinite lifetime. We have

THEOREM 1. *([V], [G], [H1]) Suppose that M is complete and that the Ricci curvature satisfies the integral condition*

$$\int_1^\infty \kappa_1(r)^{-1} dr = \infty,$$

Then M is stochastically complete.

Intuition: Since e is the explosion time of the radial process, M is stochastically complete if the Brownian motion does not wander to infinity too fast. The rate of escape of Brownian motion is bounded from above by a lower bound of the Ricci curvature, therefore a lower bound of the rate of growth of the Ricci curvature should give the stochastic completeness.

Before using the comparison theorem we need an important observation [K]: The function $r(x)$ is not a smooth function on M, but we have

$$dR_t = d\beta_t + \frac{m(R_t, \Theta_t)}{2} dt - dL_t.$$

Here L_t is a local time (positive continuous additive functional) on the cut-locus of the fixed point O. Compare the radial process R with the radial process R^* on the radially symmetric manifold with metric $ds^2 = dr^2 + G_1(r)^2 d\theta^2$:

$$dR_t^* = d\beta_t + \frac{n-1}{2} \frac{G_1'(R_t^*)}{G_1(R_t^*)} dt.$$

Noting that $L_t \geq 0$ for all $t \geq 0$ and $m(r, \theta) \leq (n-1) G_1'(r) G_1(r)^{-1}$, we can use the comparison for stochastic differential equations and obtain $R_t^* \geq R_t$. Thus R_t cannot blow up before R_t^*, i.e., $e^* \leq e$. The condition for explosion of the one dimensional process R^* is known exactly ([IW, p.362]:

$$I(G_1) = \int_c^\infty G_1(r)^{1-n} dr \int_c^r G_1(s)^{n-1} ds = \infty$$

A simple calculation shows

$$I(G_1) \geq c_1 \int_c^\infty \kappa_1(r)^{-1} dr - c_2.$$

The theorem follows. A different (semi-probabilistic) proof of this result can be found in [V], which chronologically precedes the independent proof above, given in [H1]. In [G], an even better integral condition was found, which seems to defy any probabilistic interpretation.

We now turn to another related problem. Let $C_0(M)$ be the space of continuous functions on M which vanish at infinity. The heat semigroup P_t associated with the Brownian motion is said to have the Feller property (or C_0 property) if $P_t C_0(M) \subset C_0(M)$, i.e., the function space $C_0(M)$ is invariant under P_t. We have

THEOREM 2. *([H1]) Suppose that M is complete and that the Ricci curvature satisfies the condition*

$$\int_1^\infty \kappa_1(r)^{-1} dr = \infty.$$

Then the heat semigroup has the C_0 property.

Note that we have the same condition as in the previous theorem.

Intuition: Again in this case, we need to control the rate of escape of the Brownian motion. But this case is not as obvious as the previous case. We first verify (an easy exercise) that the C_0 property is equivalent to the following statement:

For every compact set $K \subset M$ and every $t > 0$ we have

$$\lim_{r(x)\to\infty} P_x \{T_K \leq t\} = 0.$$

This means that if we start the Brownian motion from very far away, then in a finite amount of time t, it should have very small probability of coming back to a fixed compact set K. This property can be guaranteed if the Brownian motion does not travel too fast.

The proof of this theorem runs as follows. Take K to be the ball of radius R centered at O. We define a sequence of random times $\tau_1, \tau_2, \ldots$. We first draw concentric spheres $S_{r(x)}, S_{r(x)-1}, \ldots$ of radius $r(x), r(x)-1, \ldots$ with the center at O. Let τ_1 be the first exit time of the ball of radius one centered at x. Then we wait until the Brownian path reaches the sphere $S_{r(x)-1}$. Let τ_2 be the first exit time of the ball of radius one centered at the point where the path first arrives at $S_{r(x)-1}$. Then we wait again until the path reaches the sphere $S_{r(x)-2}, \ldots$, etc. Since we have a lower bound on the Ricci curvature on each small ball of radius one, namely $-L_n = -\kappa_1(r(x) - n + 1)$, on the nth ball, the portion of the path inside this ball cannot exit too fast. In other words, τ_n cannot be too small. A quantitative statement of this fact is the following estimate (see [HM]):

$$P_x\left\{\tau_n \le c_1 L_n^{-1}\right\} \le e^{-c_2 L_n},$$

where c_1, c_2 are universal constants.

Clearly we have

$$T_K \ge \tau_1 + \tau_2 + \cdots \tau_{r(x)-R}.$$

Now fix $m = m(x)$ such that

$$\int_{r(x)-m+1}^{r(x)+1} \kappa_1(r)^{-1} dr \le c_1 \sum_{n=1}^{m-1} L_n^{-1} \le t \le c_1 \sum_{n=1}^{m} L_n^{-1} \le \int_{r(x)-m}^{r(x)+1} \kappa_1(r)^{-1} dr.$$

By the condition on κ_1 in the theorem, we have $m \le r(x) - R$ for sufficiently large $r(x)$; hence

$$\begin{aligned} P_x\{T_K \le t\} &\le P_x\left\{\sum_{n=1}^{m} \tau_n \le c_1 \sum_{n=1}^{m} L_n^{-1}\right\} \\ &\le \sum_{n=1}^{m} P_x\left\{\tau_n \le c_1 L_n^{-1}\right\} \\ &\le \int_{r(x)-m}^{r(x)+1} e^{-c_2\kappa_1(r)} dr. \end{aligned}$$

Thus we have

$$P_x\{T_K \le t\} \le c_3(1+t)\, e^{-c_2\kappa_1(r(x)-m)}.$$

We may assume without loss of generality that $\kappa_1(r) \uparrow \infty$ as $r \uparrow \infty$. Now

$\int_1^\infty \kappa_1(r)^{-1}dr = \infty$ implies that $r(x) - m \to \infty$. Thus the right-hand side goes to zero and the theorem is proved.

Our next topic is harmonic functions on complete Riemannian manifolds. A question often asked about harmonic functions is: Are there non-constant harmonic functions on M? From the martingale characterization of Brownian motion, we see that if f is harmonic, then $f(X_t)$ is a continuous local martingale. This fact gives the crucial link from harmonic functions to Brownian motion. Generally, the more recurrent Brownian motion is, the less likely it is that M has non-constant harmonic functions. A well-known result of [Y] is

THEOREM 3. *On a complete manifold of nonnegative Ricci curvature, every nonnegative harmonic function is a constant.*

A weaker form of this theorem (assuming the harmonic functions to be of sublinear growth) can be proved by Brownian motion (see [DG], apparently unaware of Yau's work, and [S]). You may appreciate the following proof more if you ever attempted to study Yau's original proof. The probabilistic proof goes as follows. Since $f(X_t)$ is a continuous local martingale, we have

$$E_x\left\{f(X_t)^2\right\} \geq E_x\left\{\int_0^t |\nabla f|^2(X_s)\,ds\right\}.$$

Because f has sublinear growth, i.e., for any $\epsilon > 0$, there is a K such that $f(x)^2 \leq K^2 + \epsilon^2 r(x)^2$, we have

$$E_x\left\{|f(X_t)|^2\right\} - K^2 \leq \epsilon^2 E_x\left\{r(X_t)^2\right\} \leq \epsilon^2 E_x\left\{|W_t|^2\right\} \leq n\epsilon^2 t.$$

Here W is an n-dimensional Euclidean Brownian motion. The second inequality comes from the fact that the Ricci curvature is nonnegative, so that the Riemannian Brownian motion cannot escape faster than the Euclidean Brownian motion. On the other hand, using Itô's formula on $|\nabla f|^2$ and a little differential geometry, we have

$$|\nabla f|^2(X_t) \geq |\nabla f|^2(x) + \text{ local martingale } + \int_0^t \text{Ric}(\nabla f, \nabla f)\, ds.$$

Hence

$$K^2 + n\epsilon^2 t \geq E_x\left\{|f(X_t)|^2\right\} \geq E_x\left\{\int_0^t |\nabla f|^2(X_s)\, ds\right\} \geq |\nabla f|^2(x)\, t.$$

We conclude from this inequality that $\nabla f(x) = 0$. Therefore f is a constant.

The above proof does not work if we only assume that f is nonnegative. But in dimension 2 we have a probabilistic proof of Yau's result. Note that in this case, the result also follows from the uniformization theorem from Riemann surface theory. If you do not want to study this theory and you are following what we have said so far, you may want to see the following proof (for dimension 2).

First of all, we note that the Brownian motion in this case is recurrent, meaning that it can never leave any fixed bounded open set forever. This is because the Ricci curvature is nonnegative, and the Brownian motion X_t cannot travel farther away then the standard two-dimensional Euclidean Brownian motion, which is recurrent. Now $f(X_t)$ is a positive local martingale; it is therefore a positive supermartingale. It follows from the martingale theory that the limit $\lim_{t\to\infty} f(X_t) = Z$ exists. Now for any constant a, the function $x \mapsto P_x\{Z < a\}$ is a bounded harmonic function. Therefore it must be a constant by what we have shown before. It follows that Z must be a constant, say c. Now the recurrence of X_t forces f to be everywhere equal to c, which completes the proof.

If we assume that M is sufficiently transient then we can prove there are many bounded, non-constant harmonic functions. In a sense, a very transient manifold behaves like a bounded domain. We have the following theorem:

THEOREM 4. *([HM]) Let M be a Cartan-Hadamard manifold which satisfies the following conditions: (a)* $\text{Sect}(x) \leq -\alpha(\alpha-1)r(x)^{-2}$; *(b)* $\text{Ric}(x) \geq -L^2 r(x)^{2\beta}$; *(c) $\alpha(1-\beta) > 2$. Then there are non-constant bounded harmonic functions on M.*

The idea of the proof is to show that the angular part Θ_t has a nontrivial limit. Once we prove that $\lim_{t\to\infty} \Theta_t = \Theta_\infty$ exists and defines a nontrivial random variable on the boundary S^{n-1} at infinity, then for any bounded function f on S^{n-1}, the function $u(x) = E_x\{f(\Theta_\infty)\}$ is a bounded harmonic function on M.

Again this time, we map the Brownian motion to the tangent plane at O by the exponential map. By a theorem from differential geometry, a Cartan-Hadamard manifold (i.e., simply connected with nonpositive sectional curvature) is homeomorphic to the tangent plane, so this time on the tangent plane, we can see the global picture of Brownian motion. To show that each Brownian path approaches a limiting direction, we have to make sure that (i) Brownian motion goes to infinity at a certain rate; and (ii) Brownian motion does not swing too much. There is an obvious conflict between (i) and (ii); the former wants Brownian motion to move fast, the latter wants it to move slowly. This conflict accounts for the dependence of the upper bound and the lower bound in the theorem.

Let us sketch the proof of Theorem 4. We look at the successive lengths of time τ_n that Brownian motion takes to go distance one. Let $s_n = \tau_1 + \ldots + \tau_n$. The estimate we have used before on τ, together with a Borel-Cantelli argument, shows that $s_n \geq$const. $n^{1-\beta}$. Since the sectional curvature is assumed to be nonpositive, X_t travels at least as fast as Euclidean Brownian motion, i.e., $r(X_t) \geq t^{1/2-\epsilon}$. Hence $r(X_{s_n}) \geq n^{(1-\beta)(1/2-\epsilon)}$. Thus the angular oscillation during the time interval $[s_{n-1}, s_n]$ is at most

$$\Delta_n\theta \leq \text{ const. } n^{1+\lambda}, \qquad \lambda = \alpha(1-\beta)(1/2-\epsilon).$$

[Here we need a comparison theorem from differential geometry and the upper bound of the sectional curvature to estimate the angle bounded by two geodesics.] The condition in the theorem says that $\lambda > 0$. Hence $\sum_{n=1}^{\infty} \Delta_n\theta$ converges and the limiting angle exists.

A more refined argument shows that the limiting random variable charges every open set of S^{n-1}, which shows that the limiting angle Θ is a nontrivial random variable. The details of this proof are contained in [HM].

Further applications of probabilistic diffusion theory to geometric problems can be found in [H1] to [H6].

Note. The material presented here is based on the talk I gave at the 18th Conference on Stochastic Processes and Their Applications, held at Madison, Wisconsin from June 25 to July 1, 1989. Many people expressed the wish to see my notes in print. I must point out that the material discussed here reflects heavily my research interest in the subject, and many interesting results by other people in the same direction are not touched upon.

References

[DG] Debiard, A., Gaveau, B., and Mazet, E., Théorèmes des comparaison en géométrie Riemannienne, Publ. RIMS, Kyoto Univ., Vol. 12 (1976), 391–425.

[G] Grigor'yan, A.A., On the stochastic completeness of complete manifolds, Soviet Math. Dokl., Vol. 34 (1987), No.2, 310–313.

[GW] Greene, R.E. and Wu, H., *Function Theory on Manifolds which Possesses a Pole*, Lecture Notes in Math., No. 699, Springer-Verlag, Berlin, Heidelberg, New York, 1979.

[H1] Hsu, P., Heat semigroup on a complete Riemannian manifold, to appear in Ann. Prob.

[H2] Hsu, P., Heat Kernel on noncomplete Riemannian manifolds, to appear in Indiana Math. J.

[H3] Hsu, P. Short-time asymptotics of the heat kernel on concave boundary, to appear in Siam J. Math. Anal.

[H4] Hsu, P., Brownian motion on Riemannian manifolds, Contem. Math., No. 37(1987): *Geometry of Random Motions*, edited by R. Durrett and M. Pinsky,

95–104.

[H5] Hsu, P., Brownian bridge on complete Riemannian manifolds, to appear in Prob. Theory Rel. Fields.

[H6] Hsu, P., Brownian motion and the Atiyah-Singer index theorem, preprint.

[HM] Hsu, P. and March, P., The limiting angle of certain Riemannian Brownian motions, Comm. on Pure and Appl. Math., XXXVIII (1985), 755–768.

[IW] Ikeda, N. and Watanabe, S., *Stochastic Differential Equations and Diffusion Processes*, North-Holland/Kodansha, Amsterdam, Oxford, New York, Tokyo, 1981.

[K] Kendall, W.S., The radial part of Brownian motion on a manifold: a semimartingale property, Ann. Prob., Vol. 15 (1987), No. 4, 1491–1500.

[S] Stafford, S., A probabilistic proof of S.-Y. Cheng's Liouville theorem, preprint.

[V] Varopoulos, N.T., Potential theory and diffusion on Riemannian manifolds, *Conference on Harmonic Analysis in Honor of Antoni Zygmund*, Vol. II, edited by W. Beckner, A.P. Calderón, R. Fefferman, and P.W. Jones, Wadsworth, Inc., Belmont, CA, 1981, 821–837.

[Y] Yau, S.T., Harmonic functions on Riemannian manifolds, Comm. Pure and Appl. Math., XXVIII (1975), 201–228.

Pei Hsu
Department of Mathematics
Northwestern University
Evanston, IL 60208

PROBABILISTIC METHODS IN SCHRÖDINGER EQUATIONS

by

ZHIMING MA

RENMING SONG

1. Introduction.

It is well known that for the time homogeneous Schrödinger equation

$$\left(\frac{\Delta}{2}+q\right)u=0$$

where Δ is the Laplacian on a domain $D \subset R^d$ and q a Borel function on D, there are typically four types of boundary value problems:

I. The Dirichlet problem (or The first boundary value problem):

$$\begin{cases} \left(\frac{\Delta}{2}+q\right)u=0 & \text{in } D \\ u|_{\partial D}=f. \end{cases} \tag{1.1}$$

II. The Neumann problem (or The second boundary value problem):

$$\begin{cases} \left(\frac{\Delta}{2}+q\right)u=0 & \text{in } D \\ \frac{\partial u}{\partial n}=f & \text{on } \partial D \end{cases} \tag{1.2}$$

where $\partial/\partial n$ is the outward normal derivative and D is a sufficiently smooth domain.

III. The third boundary value problem:

$$\begin{cases} \left(\frac{\Delta}{2}+q\right)u=0 & \text{in } D \\ \frac{\partial u}{\partial n}+cu+f=0 & \text{on } \partial D \end{cases} \tag{1.3}$$

where c is a Borel function defined on ∂D and D is a sufficiently smooth domain.

IV. The mixed boundary value problem:

$$(1.4) \qquad \begin{cases} \left(\frac{\Delta}{2} + q\right) u = 0 & \text{in } D \\ \frac{\partial u}{\partial n}|_{I_2} = g_2 \\ u|_{I_1} = g_1 \end{cases}$$

where I_1 is a closed subset of ∂D and I_2 is the complement of I_1 in ∂D, g_1 and g_2 are Borel functions defined respectively on I_1 and I_2.

In 1981, Chung and Rao[**13**], by using probabilistic methods, solved the Dirichlet problem under the assumption that the underlying Schrödinger operator with Dirichlet boundary condition admits only negative eigenvalues.

In 1984, Hsu[**19**] used the reflecting Brownian motion to tackle the Neumann boundary value problem and got a probabilistic solution of the problem under a condition similar to the one assumed by Chung and Rao, i. e., the underlying Schrödinger operator with the Neumann boundary condition admits only negative eigenvalues.

In 1985, during his visit to China, Chung posed several open problems related to the probabilistic treatment of the Schrödinger equation. Among them the following are important:

(a) Is it possible to solve the Dirichlet problem probabilistically without assuming the condition that all the eigenvalues are negative?

(b) Is it possible to solve the Neumann problem without assuming the condition that all the eigenvalues are negative?

(c) Is it possible to give a probabilistic treatment of the mixed boundary value problem?

The above three open problems have been solved respectively in [**22**], [**27**],[**24**] and [**23**]. Two slightly different probabilistic approaches to the third boundary value problem were obtained independently by Papanicolaou[**29**] and Song[**33**].

In this survey we present an approach which solves the three open problems simultaneously. In fact, we are going to give a probabilistic treatment of the following boundary value problem:

$$(1.5) \qquad \begin{cases} \left(\frac{\Delta}{2} + \mu I_D\right) u + \nu I_D = 0 & \text{in } D \\ u|_{I_1} = g \\ \frac{\partial u}{\partial n}|_{I_2} - 2u\mu I_{I_2} = 2\nu I_{I_2} \end{cases}$$

where μ and ν are certain Radon measures, D is a bounded domain in R^d with $D = G - B$ where G is a bounded C^3 domain in R^d and B a closed subset of $\overline{G}$,

$I_1 = \partial D \cap B$, $I_2 = \partial D - B$, g is a Borel function defined on I_1. In our treatment the underlying Schrödinger operator can admit positive eigenvalues.

It is clear that the boundary value problem (1.5) includes the above four types of boundary value problems as special cases. The basic tools of this paper are Dirichlet form theory and stochastic calculus.

As an application of the probabilistic treatment of the boundary value problem (1.5) we will give a probabilistic proof of the existence of solutions to the following semilinear boundary value problem:

$$\begin{cases} -\left(\frac{\Delta}{2} + \mu I_D\right) u + f_1(u) = \nu I_D & \text{in } D \\ u|_{I_1} = g \\ \frac{\partial u}{\partial n}|_{I_2} - 2u\mu I_{I_2} + 2f_2(u) = 2\nu I_{I_2} \end{cases} \tag{1.6}$$

where f_1 and f_2 are continuously differentiable functions on R^1.

This paper is a summary of **[22]**, **[23]**, **[24]**, **[25]**, **[34]**, **[35]** and **[6]** with the results generalized and the arguments streamlined.

In order to keep this survey to a reasonable length, the results given in this paper are not stated in their most general forms. In fact, the assumption on the potential term μ can be weakened considerably. The interested readers can see **[7]** and **[8]** for details.

The authors would like to express their thanks to Dr. J. Glover for his help. His suggestion of a summary of the recent results on the probabilistic treatments of Schrödinger equations was the starting point of this survey and his advice led to numerous improvements of an earlier version of this paper.

2. The Reflecting Brownian Motion.

We shall always assume that G is a bounded C^3 domain in R^d $(d \geq 2)$, (X_t, P_x) the reflecting Brownian motion on $\overline{G}$ defined in **[19]** and **[31]**, L_t the boundary local time of X_t, $p(t,x,y)$ the transition density of X_t. We shall sometimes write

$$P_t\mu(x) = \int_{\overline{G}} p(t,x,y)\,\mu(dy)$$
$$G_\alpha\mu(x) = \int_0^\infty e^{-\alpha t} P_t\mu(x)\,dt$$

provided the right hand sides make sense.

Set

$$C^2(\overline{G}) = \{u = f|_{\overline{G}} : f \in C_0^2(R^d)\},$$
$$H^1(\overline{G}) = \{u \in L^2(\overline{G}) : \frac{\partial u}{\partial x^i} \in L^2(\overline{G}), 1 \leq i \leq d\}.$$

The derivatives are generally taken in the distributional sense unless otherwise stated. For u, $v \in H^1(\overline{G})$, define

$$\mathcal{E}(u,v) = \frac{1}{2}\int_G \nabla u \nabla v \, dx$$

and

$$\begin{aligned}\mathcal{E}_\alpha(u,v) &= \mathcal{E}(u,v) + \alpha(u,v)\\ &= \frac{1}{2}\int_G \nabla u \nabla v \, dx + \alpha \int_G uv \, dx, \qquad \alpha > 0\end{aligned}$$

Then $(\mathcal{E}, H^1(\overline{G}))$ is a regular Dirichlet form with $C^2(\overline{G})$ as its core.

Using the Tanaka formula (see **[19]**) and Lemma 1.3.4 of **[16]** it is easy to see that $(\mathcal{E}, H^1(\overline{G}))$ is the Dirichlet form corresponding to the reflecting Brownian motion X_t.

2.1 DEFINITION[**16**]. *A positive finite measure μ on $\overline{G}$ is said to be of finite energy integral, denoted by $\mu \in S_0(\overline{G})$, if there exists a constant $C > 0$ such that*

$$\int_G |f(x)|\,\mu(dx) \le C \cdot \sqrt{\mathcal{E}_1(f,f)} \qquad \forall f \in C^2(\overline{G}).$$

By Riesz's representation theorem, a finite measure $\mu \in S_0(\overline{G})$ iff for each $\alpha > 0$ there exists an element $U_\alpha\mu \in H^1(\overline{G})$ such that

$$\mathcal{E}_\alpha(U_\alpha\mu, \phi) = \int_G \phi(x)\,\mu(dx), \qquad \forall \phi \in C^2(\overline{G}).$$

It follows from Chapter 5 of **[16]** that every $\mu \in S_0(\overline{G})$ is associated with a unique continuous additive functional A_t of X_t such that $E_x \int_0^\infty e^{-t}\,dA_t$ is a quasicontinuous version of $U_1\mu$.

The following result gives a necessary and sufficient condition for a measure $\mu \in S_0(\overline{G})$.

2.2 PROPOSITION. *Let μ be a finite measure on $\overline{G}$. Then $\mu \in S_0(\overline{G})$ iff for some $\alpha > 0$, $G_\alpha\mu \in L^2(\overline{G})$ and*

$$\lim_{t\downarrow 0}\left(\frac{1}{t}(G_\alpha\mu - e^{-\alpha t}P_t(G_\alpha\mu)), G_\alpha\mu\right) < \infty.$$

The proof is very easy and therefore is omitted here.

2.3 DEFINITION. *A signed Radon measure μ on R^d is said to be in the generalized Kato class, denoted by $\mu \in GK^d$, if*

$$\lim_{r\downarrow 0}\sup_{x\in R^d}\int_{B(x,r)} k(x,y)\,|\mu|(dy) = 0$$

where

$$k(x,y) = \begin{cases} -\log|x-y|, & d=2 \\ |x-y|^{2-d}, & d \geq 3. \end{cases}$$

In the sequel we shall use $GK^d(\overline{G})$ *to denote the class of measures in* GK^d *such that* $\mu(R^d - \overline{G}) = 0$.

2.4 PROPOSITION. *Let* μ *be a finite measure on* $\overline{G}$. *Then* $\mu \in GK^d(\overline{G})$ *iff*

$$\lim_{t\downarrow 0} \sup_{x\in\overline{G}} \int_0^t \int_{\overline{G}} p(s,x,y)\,\mu(dy)ds = 0$$

The special case of $\mu(dx) = q(x)dx$ is proved in [**19**] and the general case goes along the same line.

2.5 THEOREM. *A measure* $\mu \in GK^d(\overline{G})$ *iff the following three conditions are satisfied:*

(1) $|\mu| \in S_0(\overline{G})$;

(2) *the continuous additive functional* A_t *of* X_t *associated with* μ *can be defined without exceptional set;*

(3) $\lim_{t\downarrow 0}\sup_{x\in\overline{G}} E_x|A|_t = 0$ *with* $|A|_t = A_t^+ + A_t^-$.

PROOF: Without loss of generality we may assume that μ is a positive measure. Suppose first that μ satisfies (1)—(3). It follows from Lemma 5.1.4 of [**16**] that for any $h, f \in \mathcal{B}^+$

$$\left(h, E_\cdot \int_0^t f(X_s)\,dA_s\right) = \left(h, \int_0^t \int_{\overline{G}} p(s,\cdot,y) f(y)\,\mu(dy)ds\right).$$

Consequently for fixed f we have

$$E_x \int_0^t f(X_s)\,dA_s = \int_0^t \int_{\overline{G}} p(s,x,y) f(y)\,\mu(dy)ds, \qquad \text{a.e. dx.}$$

Multiplying both sides of the above equality by $p(\epsilon, x_0, x)$ and integrating over x, we get

$$E_{x_0} \int_\epsilon^t f(X_s)\,dA_s = \int_\epsilon^t \int_{\overline{G}} p(s,x_0,x) f(y)\,\mu(dy)ds.$$

In particular, since x_0 is an arbitrary point, we get

$$E_x A_t = \int_0^t P_s\mu(x)\,ds, \qquad \forall x \in \overline{G}.$$

Therefore by (3) we have

$$\lim_{t\downarrow 0} \sup_{x\in\overline{G}} \int_0^t P_s\mu(x)\,ds = 0$$

which is equivalent to $\mu \in GK^d(\overline{G})$.

To prove the converse, suppose now that $\mu \in GK^d(\overline{G})$. Since

$$\lim_{t\downarrow 0} \sup_{x\in\overline{G}} \int_0^t \int_{\overline{G}} p(s,x,y)\,|\mu|(dy)ds = 0,$$

it is easy to show that

$$\sup_{x\in\overline{G}} \int_0^\infty e^{-s} P_s\mu(x)\,ds := M < \infty;$$

therefore

$$\begin{aligned}\lim_{t\downarrow 0} \frac{1}{t}\left(G_1\mu - e^{-t}P_t(G_1\mu), G_1\mu\right) &\le \lim_{t\downarrow 0} \frac{M}{t} \int_{\overline{G}} \int_0^t e^{-s} P_s\mu(x)\,dsdx \\ &\le M\mu(\overline{G}) < \infty.\end{aligned}$$

Applying Proposition 2.2 we conclude that $\mu \in S_0(\overline{G})$. Let A_t be the continuous additive functional of X_t associated with μ. Then

$$\begin{aligned}E_x \int_0^\infty e^{-s}\,dA_s &= \int_0^\infty e^{-s} \int_{\overline{G}} p(s,x,y)\,\mu(dy)ds \\ &= G_1\mu(x), \qquad \text{a.e. dx}\end{aligned}$$

On the other hand, it is easy to verify that $G_1\mu$ is a bounded continuous 1-excessive function in the sense of [**9**], consequently there exists a unique continuous additive functional $\overline{A}_t$ without exceptional set such that

$$E_x \int_0^\infty e^{-s}\,d\overline{A}_s = G_1\mu(x).$$

By uniqueness A_t must be equivalent to $\overline{A}_t$, which means that A_t can be defined without exceptional set. Repeating the first part of the proof we can show that (3) is true.

Let us denote by σ the "area"measure on ∂G. Then the following result is well known:

2.6 PROPOSITION. *L_t is the positive continuous additive functional of X_t associated with the measure σ.*

2.7 PROPOSITION. *Let f be a bounded Borel function on ∂G. Define a measure μ by $\mu(dx) = f(x)\sigma(dx)$. Then $\mu \in GK^d(\overline{G})$.*

PROOF: Using the trace theorem (see [**1**]) we can show that μ satisfies the conditions (1)—(3) of Theorem 2.5, therefore $\mu \in GK^d(\overline{G})$. For details, see [**25**].

3. The Mixed Barrier Brownian Motion.

Let D be a bounded domain in R^d. Assume that $D = G - B$ where B is a closed subset of $\overline{G}$. Put $I_1 = \partial D \cap B$, $I_2 = \partial D - B$ and $\tilde{D} = D \cup I_2 := \overline{G} - B$. Adjoin an extra point Δ to $\tilde{D}$ and regard $\tilde{D}_\Delta = \tilde{D} \cap \{\Delta\}$ as the one-point compactification of $\tilde{D}$.

Set

$$\dot{\tau}(\omega) = \inf\{t \geq 0 : X_t(\omega) \in B\}$$
$$\tau(\omega) = \inf\{t > 0 : X_t \in B\}$$

and define

$$X_t^0(\omega) = \begin{cases} X_t(\omega), & 0 \leq t < \dot{\tau}(\omega) \\ \Delta, & t \geq \dot{\tau}(\omega). \end{cases}$$

Then $(\Omega, \mathcal{F}, X_t^0, P_x)$ is a standard process with state space $\tilde{D}$ and life time τ. Moreover, X_t^0 is a Hunt process. Intuitively speaking, X_t^0 is a Brownian motion on $\tilde{D}$ with I_1 as its absorbing barrier and I_2 as its reflecting barrier. We call X_t^0 the mixed barrier Brownian motion on $\tilde{D}$.

For $f \in \mathcal{B}(\tilde{D})$, define

$$P_t^0 f(x) = E_x f(X_t^0) := E_x\left(f(X_t) I_{\{t<\tau\}}\right), \qquad x \in \tilde{D},$$

provided the right hand side makes sense. Restricted to $L^2(\tilde{D})$, P_t^0 is nothing but the strongly continuous Markovian semigroup associated with X_t^0.

Invoking the results of chapter 1 of [**19**] or the corresponding results in [**31**] and using an argument similar to that used in Theorem 2.4.3 of [**30**] we can get the following result:

3.1 THEOREM. *$(P_t^0)_{t>0}$ admits an integral kernel $p^0(t,x,y)$ defined on $(0,\infty) \times \tilde{D} \times \tilde{D}$ which satisfies the following properties:*

(1)

$$P_t^0 f(x) = \int_{\tilde{D}} p^0(t,x,y) f(y)\, dy \qquad \forall f \in \mathcal{B}^+(\tilde{D}), x \in \tilde{D};$$

(2) *$p^0(t,x,y)$ is jointly continuous on $(0,\infty) \times \tilde{D} \times \tilde{D}$;*

(3)

$$p^0(t+s,x,y) = \int_{\tilde{D}} p^0(t,x,z) p^0(s,z,y)\, dz$$

(4) *$p^0(t,x,y)$ is symmetric in x and y;*

(5)

$$0 < p^0(t,x,y) \leq p(t,x,y) \leq C_1 + C_2 t^{-d/2}$$

where $p(t,x,y)$ is the transition density function of X_t on $\bar{G}$, C_1 and C_2 are constants.

(6) *Let x_0 (or y_0) be a regular point of I_1 (i.e., $P_{x_0}(\tau = 0) = 1$), $t_0 > 0$, then $\lim_{n\to\infty} p^0(t_n, x_n, y_n) = 0$ whenever (t_n, x_n, y_n) converges to (t_0, x_0, y_0).*

PROOF: For details, see **[23]**.

Set

$$C_0^2(\tilde{D}) = \{u = f|_{\tilde{D}} : f \in C_0^2(R^d) \text{ with compact support in } R^d - B\},$$

and let $H_0^1(\tilde{D})$ be the closure of $C_0^2(\tilde{D})$ with respect to the following norm

$$\|u\|^2 = \int_{\tilde{D}} u^2\,dx + \frac{1}{2}\int_{\tilde{D}} (\nabla u)^2\,dx.$$

Then $(\mathcal{E}, H_0^1(\tilde{D}))$ is a regular Dirichlet form with $C_0^2(\tilde{D})$ as its core.

3.2 PROPOSITION. *$(\mathcal{E}, H_0^1(\tilde{D}))$ is the Dirichlet form corresponding to the mixed barrier Brownian motion X_t^0.*

PROOF: The proof of this result is quite easy and therefore is omitted here.

Set

$$S_0(\tilde{D}) = \{\mu \in S_0(\overline{G}) : \mu(\overline{G} - \tilde{D}) = 0\}$$
$$GK^d(\tilde{D}) = \{\mu \in GK^d(\overline{G}) : \mu(\overline{G} - \tilde{D}) = 0\}$$

Let $\mu \in S_0(\tilde{D})$. Then μ, being a measure of finite energy integral with respect to $(\mathcal{E}, H^1(\overline{G}))$, is associated with a unique continuous additive functional A_t of X_t. μ is also a measure of finite energy integral with respect to $(\mathcal{E}, H_0^1(\tilde{D}))$, therefore μ is also associated with a unique continuous additive functional A_t^0 of X_t^0. From Oshima**[28]** we know that $A_t^0 = A_{t\wedge\tau}$. In what follows we shall rely heavily on this fact.

4. The Formulation of the Mixed Boundary Value Problem.

From now on we are going to assume that $\mu, \nu \in GK^d(\tilde{D})$ are fixed measures and that A_t and B_t are the continuous additive functionals of X_t associated respectively with μ and ν. We are also going to assume that g is a bounded Borel function on I_1.

In this section we are going to explain what we mean by saying that a function u is a solution to the following boundary value problem:

$$(4.1) \qquad \begin{cases} \left(\dfrac{\Delta}{2} + \mu I_D\right) u + \nu I_D = 0 \\ u|_{I_1} = g \\ \dfrac{\partial u}{\partial n}|_{I_2} - 2u\mu I_{I_2} = 2\nu I_{I_2}. \end{cases}$$

But before doing that we must define $\partial u/\partial n|_{I_2}$, the normal derivative of a function u on I_2.

Set

$$H^1_{loc}(\tilde{D}) = \{u \in \mathcal{B}(\tilde{D}) : uh \in H^1_0(\tilde{D}) \text{ for all } h \in C^2_0(\tilde{D})\}$$
$$GK^d_{loc}(\tilde{D}) = \{\mu : h(x)\mu(dx) \in GK^d(\tilde{D}) \text{ for all } h \in C_0(\tilde{D})\}$$

4.1 DEFINITION. *Let $\bar{\mu} \in GK^d_{loc}(\tilde{D})$ be such that $|\bar{\mu}|(D) = 0$ and let $u \in L^1_{loc}(\tilde{D})$ be such that u admits a quasicontinuous modification $\tilde{u}$, Δu is a Radon measure on $\tilde{D}$ and $\tilde{u}|_{I_2} \in L^1_{loc}(I_2, \sigma)$. If*

$$\int_{\tilde{D}} h\,(\Delta u)(dx) - \int_{\tilde{D}} u\Delta h\, dx = \int_{I_2} h\,\bar{\mu}(dx) - \int_{I_2} \frac{\partial u}{\partial n}\tilde{u}\,\sigma(dx), \qquad \forall h \in C^2_0(\tilde{D})$$

then $\bar{\mu}$ is called a (weak) normal derivative of u on I_2, denoted by $\partial u/\partial n|_{I_2}$.

It is clear that $\partial u/\partial n|_{I_2}$ is uniquely determined, and if $u \in C^1(\tilde{D}) \cap C^2(D)$ with $\Delta u \in L^1(\tilde{D})$, then $\partial u/\partial n|_{I_2}$ coincides with the classical normal derivative. From now on, the quasicontinuous modification of a function u, so long as it exists, will be denoted by $\tilde{u}$.

The following result is quite easy to prove.

4.2 PROPOSITION. *Let $u \in H^1_{loc}(\tilde{D})$ be such that Δu is a Radon measure on $\tilde{D}$ and $\tilde{u}|_{I_2} \in L^1_{loc}(I_2, \sigma)$. Then $\bar{\mu} \in GK^d_{loc}(\tilde{D})$ with $|\bar{\mu}|(D) = 0$ is a normal derivative of u on I_2 iff*

$$\int_{\tilde{D}} \nabla u \cdot \nabla h\, dx + \int_{\tilde{D}} h\,(\Delta u)(dx) = \int_{I_2} h\,\bar{\mu}(dx), \qquad \forall h \in C^2_0(\tilde{D}).$$

In what follows we shall give a probabilistic characterization of the normal derivative $\partial u/\partial n|_{I_2}$. To this end let us fix a smooth measure $\gamma \in GK^d_{loc}(\tilde{D})$ and we denote by C_t the continuous additive functional of X_t associated with γ.

For $u \in L^1(\tilde{D})$ we define

$$Q_t(u,\gamma)(x) = E_x\left(u(X_t)I_{\{t<\tau\}} + C_{t\wedge\tau}\right)$$

provided the right hand side makes sense. It is easy to prove that if $\gamma \in GK^d(\tilde{D})$ then $Q_t(u,\gamma) \in bC(\tilde{D})$ for each $t > 0$.

4.3 PROPOSITION. *Let $u \in H^1_0(\tilde{D})$. Then the following two assertions are equivalent.*

(1) $u(x) = Q_t(u,\gamma)(x)$, a.e.. dx, $\forall t > 0$.
(2) $\Delta u + 2\gamma I_D = 0$ and $\partial u/\partial n|_{I_2} = 2\gamma I_{I_2}$.

PROOF: (1)$\Longrightarrow$(2) Suppose that (1) holds. For $h \in C_0^2(\tilde{D})$, by Lemma 1.3.4 of [16] we have

$$\frac{1}{2}\int_{\tilde{D}} \nabla u \cdot \nabla h\, dx = \lim_{t\downarrow 0}\int_{\tilde{D}} h(x)\frac{u(x) - P_t^0 u(x)}{t}\, dx = \int_{\tilde{D}} h\,\gamma(dx).$$

In particular

$$\int_D \nabla u \cdot \nabla h\, dx = 2\int_D h\,\gamma(dx) \qquad \forall h \in C_0^2(D),$$

which implies that $\Delta u + 2\gamma I_D = 0$. It follows from Proposition 4.2 that $\partial u/\partial n|_{I_2} = 2\gamma I_{I_2}$.

(2)$\Longrightarrow$(1) Suppose that (2) holds. For $h \in C_0^2(\tilde{D})$, set

$$\xi_t = \int_{\tilde{D}} h(u - P_t^0 u)\, dx$$
$$\eta_t = \int_{\tilde{D}} h(x) E_x C_{t\wedge T}\, dx$$

It follows from Lemma 1.3.4 of [16] and (2) of this proposition that

$$\frac{d\xi_t}{dt} = \frac{d\eta_t}{dt} = \frac{1}{2}\int_{\tilde{D}} \nabla u \cdot \nabla(P_t^0 h)\, dx$$

Since $\xi_0 = \eta_0 = 0$, we get

$$\int_{\tilde{D}} h(u - P_t^0 u)\, dx = \int_{\tilde{D}} h(x) E_x C_{t\wedge T}\, dx \qquad \forall h \in C_0^2(\tilde{D}).$$

Consequently

$$u - P_t^0 u = E.C_{t\wedge T}, \qquad \text{a.e. dx.}$$

In the sequel we shall make the convention that a function u defined on $\tilde{D}$ may be automatically extended to $\overline{G}$ by setting $u(x) = 0$ for $x \notin \tilde{D}$.

4.4 LEMMA. *Let $h \in C_0^2(\tilde{D})$.*

(1) *Suppose that $u \in H^1_{loc}(\tilde{D})$. Then $uh \in H^1(\overline{G})$ and*

$$\nabla(uh) = u\nabla h + h\nabla u.$$

(2) *Suppose that $u \in H^1_{loc}(\tilde{D})$ satisfying $\Delta u \in GK^d_{loc}(\tilde{D})$. Then $\Delta(uh) \in GK^d(\tilde{D})$.*

(3) *Suppose that $u \in H^1_{loc}(\tilde{D})$ satisfying $\Delta u \in GK^d_{loc}(\tilde{D})$, and that u admits a normal derivative on I_2. Then uh admits a normal derivative on ∂G and*

$$\frac{\partial(uh)}{\partial n}|_{I_2} = \tilde{u}\frac{\partial h}{\partial n}|_{I_2} + h\frac{\partial u}{\partial n}|_{I_2}.$$

4.5 LEMMA. *Let $u \in H^1(\overline{G})$ and $v \in H_0^1(\tilde{D})$, then*

$$\int_{\tilde{D}} uv\,dx + \frac{1}{2}\int_{\tilde{D}} \nabla u \cdot \nabla v\,dx = \lim_{t\downarrow 0}\int_{\tilde{D}} v(x)\left(\frac{u(x) - E_x(e^{-t\wedge\tau}\tilde{u}(X_{t\wedge\tau}))}{t}\right)dx.$$

The proofs of the above two lemmas are straightforward.

Recall that by definition for an arbitrary Borel subset A of $\overline{G}$

$$\tau_A = \inf\{t > 0 : X_t \in A\}.$$

4.6 PROPOSITION. *Let $u \in H^1_{loc}(\tilde{D})$, $\gamma \in GK^d_{loc}(\tilde{D})$. Then the following two assertions are equivalent.*

(1) *For each relatively open subset A of $\overline{G}$ such that $A \supset B$, it holds that for a.e. $x \in \overline{G} - A$.*

$$u(x) = E_x\left(\tilde{u}(X_{t\wedge\tau_A}) + C_{t\wedge\tau_A}\right), \qquad \forall t > 0.$$

(2) $\Delta u + 2\gamma I_D = 0$ and $\partial u/\partial n|_{I_2} = 2\gamma I_{I_2}$.

PROOF: (1)$\Longrightarrow$(2) Let $v \in C_0^2(\tilde{D})$. Take a relatively open subset A of $\overline{G}$ such that $A \supset B$ and $\overline{G} - \overline{A}$ contains the support of v. Write $T = \tau_{\overline{A}}$. It follows from (1) that for a.e. $x \in \overline{G} - A$,

$$\{\tilde{u}(X_{t\wedge\tau}) - \tilde{u}(X_0) + C_{t\wedge\tau}\}_{t\geq 0}$$

is a continuous P_x-martingale. Applying the integration by parts formula with respect to semimartingales we obtain that for a.e. $x \in \overline{G} - A$

$$E_x\left(e^{-t\wedge T}\tilde{u}(X_{t\wedge T})\right) = E_x\left(\tilde{u}(X_0) - \int_0^{t\wedge T} e^{-s}\,dC_s - \int_0^{t\wedge T} e^{-s}\tilde{u}(X_s)\,ds\right).$$

Consequently

$$\lim_{t\downarrow 0}\int_{\overline{G}-\overline{A}} v(x)\frac{\tilde{u}(x) - E_x e^{-t\wedge T}\tilde{u}(X_{t\wedge T})}{t}\,dx = \int_{\overline{G}-\overline{A}} \tilde{u}v\,dx + \int_{\overline{G}-\overline{A}} v(x)\,\gamma(dx).$$

Take a function $h \in C_0^2(\tilde{D})$ such that $h|_{\overline{G}-A} \equiv 1$. Consider $H_0^1(\overline{G} - \overline{A})$ instead of $H_0^1(\tilde{D})$. By Lemma 4.5 we have

$$\begin{aligned}&\int_{\overline{G}-\overline{A}}(uh)v\,dx + \frac{1}{2}\int_{\overline{G}-\overline{A}} \nabla(uh)\nabla v\,dx\\ &= \lim_{t\downarrow 0}\int_{\overline{G}-\overline{A}} v(x)\frac{\tilde{u}h(x) - E_x e^{-t\wedge T}\tilde{u}h(X_{t\wedge T})}{t}\,dx\end{aligned}$$

Since the support of v is contained in $\overline{G}-\overline{A}$ and $h|_{\overline{G}-\overline{A}} \equiv 1$, the above two equalities show that

$$\int_{\tilde{D}} \nabla u \cdot \nabla v \, dx = 2\int_{\tilde{D}} v\,\gamma(dx), \qquad \forall v \in C_0^2(\tilde{D}).$$

Let v runs over $C_0^2(D)$, we can get that $\Delta u = 2\gamma I_D$ and consequently $\partial u/\partial n|_{I_2} = 2\gamma I_{I_2}$ by Proposition 4.2. Thus (1)$\Longrightarrow$(2) is proved.

(2)$\Longrightarrow$(1) For an arbitrary relatively open subset A of $\overline{G}$ such that $A \supset B$, take a function $h \in C_0^2(\tilde{D})$ satisfying $h|_{\overline{G}-A} \equiv 1$. By Proposition 4.3, Lemma 4.4 and Doob's stopping theorem we can get that for a.e. $x \in \overline{G}-A$,

$$u(x) = E_x\left(\tilde{u}(X_{t\wedge\tau_A}) + C_{t\wedge\tau_A}\right), \qquad \forall t > 0$$

which is (1).

The proof is now complete.

Before giving the precise formulation of problem (4.1) we must introduce the concept of a regular $\tilde{D}$-harmonic function.

4.7 DEFINITION. *A bounded continuous function u on $\tilde{D}$ is said to be a $\tilde{D}$-harmonic function if $\Delta u = 0$ and $\partial u/\partial n|_{I_2} = 0$. The totality of $\tilde{D}$-harmonic functions will be denoted by $\mathcal{H}(\tilde{D})$.*

4.8 DEFINITION. *A function $u \in \mathcal{H}(\tilde{D})$ is said to be a regular $\tilde{D}$-harmonic function, if there exists a sequence $(A_n, u_n)_{n\geq 1}$ such that*

(1) *each A_n is a relatively open subsets of $\overline{G}$, $A_n \subset \overline{A}_n \subset A_{n+1} \subset \overline{A}_{n+1} \subset \tilde{D}$ and $\bigcup A_n = \tilde{D}$.*
(2) *$u_n \in \mathcal{H}(A_n) \cap H^1_{loc}(A_n)$ and*

$$\lim_{m\to\infty} \sup_{x\in A_n} |u_m(x) - u(x)| = 0, \qquad n \geq 1.$$

The totality of regular $\tilde{D}$-harmonic functions will be denoted by $\mathcal{H}_r(\tilde{D})$.

The following two results are important for the probabilistic treatment of (4.1) in later sections and are not difficult to prove.

4.9 PROPOSITION. *Let $u \in C(\tilde{D})$. Then $u \in \mathcal{H}_r(\tilde{D})$ iff for each relatively open subset $A \subset \overline{A} \subset \tilde{D}$, it holds that*

$$u(x) = E_x u(X(t \wedge \tau_{\overline{G}-A})), \qquad x \in A.$$

4.10 PROPOSITION. *Let $h \in b\mathcal{B}(I_1)$. Define*

$$u(x) = E_x h(X_\tau), \qquad x \in \tilde{D}.$$

Then $u \in \mathcal{H}_r(\tilde{D})$.

Now we are in a position to give the precise formulation of the problem (4.1).

4.11 DEFINITION. *A bounded continuous function u on $\tilde{D}$ is said to be a solution to the problem (4.1) if*

(1) $u \in \mathcal{H}_r(\tilde{D}) \oplus H_0^1(\tilde{D})$;

(2) *u satisfies*

$$\left(\frac{\Delta}{2} + \mu I_D\right) u + \nu I_D = 0, \qquad \text{in } D$$

in the distributional sense;

(3) *u satisfies*

$$\frac{\partial u}{\partial n}|_{I_2} = 2\nu I_{I_2} + 2u\mu I_{I_2}$$

in the sense of Definition 4.1;

(4) *for any $x \in \tilde{D}$*

$$\lim_{t \uparrow \tau} u(X_t) = g(X_\tau), \qquad P_x \text{ a.s. on } \{\tau < \infty\}.$$

5. The Schrödinger Semigroup.

For any $f \in \mathcal{B}(\tilde{D})$, define

$$\begin{aligned} e^{tH} f(x) &= E_x\left(e^{A_{t\wedge\tau}} f(X_t^0)\right) \\ &= E_x\left(e^{A_t} f(X_t) I_{\{t<\tau\}}\right) \end{aligned}$$

provided the right hand side makes sense. By Khas'minskii's lemma we know that e^{tH} defines a bounded operator on $L^\infty(\tilde{D})$ at least for small t. By the additive property of A_t it is easy to see that

$$e^{(t+s)H} f = e^{tH}(e^{sH} f), \qquad \forall t, s \geq 0. \tag{5.1}$$

Therefore $(e^{tH})_{t>0}$ forms a semigroup in $L^\infty(\tilde{D})$.

In this section we are going to study some properties of this semigroup $(e^{tH})_{t>0}$. Set

$$c_t = \sup_x E_x |A|_t,$$

$$q_0(t, x, y) = \bar{q}_0(t, x, y) = p^0(t, x, y),$$

$$q_n(t, x, y) = \int_0^t \int_{\tilde{D}} p^0(s, x, z) q_{n-1}(t - s, z, y)\, \mu(dz) ds, \tag{5.2}$$

$$\bar{q}_n(t, x, y) = \int_0^t \int_{\tilde{D}} p^0(s, x, z) \bar{q}_{n-1}(t - s, z, y)\, |\mu|(dz) ds, \tag{5.3}$$

5.1 PROPOSITION. *For $t \le 1$ the following properties are satisfied.*

(1) *There exists a positive constant K such that*

$$q_n(t,x,y) \le \bar{q}_n(t,x,y) \le K^{n+1} c_t^n t^{-d/2}.$$

(2) *For any $f \in \mathcal{B}^+(\tilde{D})$.*

$$\int_{\tilde{D}} q_n(t,x,y) f(y)\, dy = \frac{1}{n!} E_x\left(A_t^n f(X_t) I_{\{t<\tau\}}\right).$$

(3)

$$\int_0^t \int_{\tilde{D}} \bar{q}_n(s,x,y)\, |\mu|(dy) ds < C_t^{n+1}.$$

(4) *$q_n(t,x,y)$ is jointly continuous on $(0,1] \times \tilde{D} \times \tilde{D}$.*

(5) *$q_n(t,x,y)$ is symmetric in x and y.*

PROOF: By Theorem 3.1, there exists a constant c such that $p^0(t,x,y) \le ct^{-d/2}$ for $0 < t \le 1$. Set $K = 2^{d/2}(1+c)$. Then all the assertions (1)—(5) hold obviously for $n = 0$. Suppose that (1)—(5) hold for $n = m-1$ and consider the case of $n = m$. (1) is obtained by splitting the integral (5.2) into two parts: from 0 to $\frac{t}{2}$ and from $\frac{t}{2}$ to t. (2) is obtained by using the additive property of A_t and the Markov property. Applying (2) we can get

$$\int_0^t \int_{\tilde{d}} \bar{q}_m(s,x,y)\, |\mu|(dy) ds$$
$$= \int_0^t \int_{\tilde{D}} \left(\int_0^s \int_{\tilde{D}} p^0(u,x,z) \bar{q}_{m-1}(s-u,z,y)\, |\mu|(dz) du \right) |\mu|(dy) ds.$$

Letting $s - u = v$, the above equality becomes

$$\int_0^t \int_{\tilde{D}} \bar{q}_m(s,x,y)\, |\mu|(dy) ds$$
$$= \int_0^t \int_{\tilde{D}} p^0(u,x,z) \left(\int_0^{t-u} \int_{\tilde{D}} \bar{q}_{m-1}(v,z,y)\, |\mu|(dy) dv \right) |\mu|(dz) du$$

from which (3) follows immediately.

We have by (1) and (3)

$$\lim_{\epsilon \downarrow 0} \sup_{\delta < t < 1-\delta} \left(\int_0^\epsilon I_{\{s<t\}} \int_{\tilde{D}} p^0(s,x,z) \bar{q}_{m-1}(t-s,z,y)\, |\mu|(dz) ds \right.$$
$$\left. \int_\epsilon^1 I_{\{t-\epsilon<s\le t\}} \int_{\tilde{D}} p^0(s,x,z) \bar{q}_{m-1}(t-s,z,y)\, |\mu|(dz) ds \right) = 0$$

and by the dominated convergence theorem, the integral

$$\int_\epsilon^1 I_{\{s \le t-\epsilon\}} \int_{\tilde{D}} p^0(s,x,z) q_{m-1}(t-s,z,y)\, \mu(dz) ds$$

is jointly continuous on $(\delta, 1-\delta) \times \tilde{D} \times \tilde{D}$ for each $\epsilon > 0$. Then (4) is proved for $n = m$. (5) is obviously valid by the expression (5.2). The proof is completed by induction.

5.2 THEOREM. *$(e^{tH})_{t>0}$ admits an integral kernel $q(t,x,y)$ which satisfies the following properties:*

(1) *$q(t,x,y)$ is jointly continuous on $(0,\infty)\times\tilde{D}\times\tilde{D}$;*

(2) *there exist constants c and β such that*

$$0 < q(t,x,y) \le ct^{-d/2}e^{\beta t}; \tag{5.4}$$

(3) *$q(t,x,y)$ is symmetric in x and y;*

(4) *for any $f \in \mathcal{B}^+(\tilde{D})$,*

$$\int_{\tilde{D}} q(t,x,y)f(y)\,dy = e^{tH}f(x); \tag{5.5}$$

(5) *for any $t,s>0$,*

$$q(t+s,x,y) = \int_{\tilde{D}} q(t,x,z)q(s,z,y)\,dz; \tag{5.6}$$

(6) *if f is bounded on $\tilde{D}$ and continuous at x, then*

$$\lim_{t\downarrow 0}\int_{\tilde{D}} q(t,x,y)f(y)\,dy = f(x).$$

PROOF: Choose $t_0 > 0$ such that $Kc_t < 1$ for $t < t_0$ where K is specified by Proposition 5.1 (1). Define $q(t,x,y) = \sum_0^\infty q_n(t,x,y)$ first for $t < t_0$ and then extend for arbitrary $t > 0$ by formula (5.6). Applying Proposition 5.1, it is easy to check that $q(t,x,y)$ satisfies (1)—(6).

The following result is a direct consequence of Theorem 5.2.

5.3 THEOREM.

(1) *For each $t > 0$, e^{tH} is a bounded operator from $L^1(\tilde{D})$ to $bC(\tilde{D})$.*

(2) *$(e^{tH})_{t>0}$ is a strongly continuous, bounded operator semigroup on $L^p(\tilde{D})$, for each $1 \le p < \infty$.*

Using Proposition 4.3 and the above theorem we can get the following result:

5.4 THEOREM. *Let $u \in L^1(\tilde{D})$ and λ be a real number. Then the following statements are equivalent:*

(1) *u is a solution to the following problem*

$$\begin{cases} \left(\dfrac{\Delta}{2} + \mu I_D\right)u = \lambda u \\ u|_{I_1} = 0 \\ \dfrac{\partial u}{\partial n}|_{I_2} - 2u\mu I_{I_2} = 0; \end{cases}$$

(2) *$e^{tH}u = e^{\lambda t}u, \qquad \forall t > 0;$*

(3) *there exists a $t > 0$ such that $e^{tH}u = e^{\lambda t}u$.*

Concerning the generator of the semigroup $(e^{tH})_{t>0}$, we have the next result.

5.5 THEOREM. *Let H be the generator of $(e^{tH})_{t>0}$ on $L^2(\tilde{D})$ and $\mathcal{D}(H)$ be the domain of H. Then*

(1) *$\mathcal{D}(H)$ is the family of functions in $H_0^1(\tilde{D})$ such that*

$$|u| \cdot |\mu| \in S_0(\tilde{D}), (\Delta/2 + \mu I_D)u \in L^2(\tilde{D}) \text{ and } \frac{\partial u}{\partial n}|_{I_2} - 2u\mu I_{I_2} = 0.$$

(2) *for $u \in \mathcal{D}(H)$,*

$$Hu = \left(\frac{\Delta}{2} + \mu I_D\right) u$$

in the distributional sense.

For the proof of Theorem 5.5 see the forthcoming paper[**26**].

The above theorem shows that H is nothing but the Schrödinger operator $\Delta/2 + \mu$ restricted to D with mixed boundary conditions $u|_{I_1} = 0$ and $\partial u/\partial n|_{I_2} - 2u\mu I_{I_2} = 0$. For this reason we shall write $(\Delta/2 + \mu)|_{\tilde{D}}$ in place of H to suggest its intuitive meaning.

By Theorems 5.2–5.5 we know that there exists a decreasing sequence $\{\lambda_i\}_{i\geq 1}$ of eigenvalues of $(\Delta/2 + \mu)|_{\tilde{D}}$ such that $\lambda_i \downarrow -\infty$ and the corresponding eigenfunctions $\{\phi_i\}_{i\geq 1}$ can be chosen so that they all belong to $bC(\tilde{D})$ and so that they form an orthornormal base of $L^2(\tilde{D})$. In the sequel we shall always assume that $\{\lambda_i\}_{i\geq 1}$ and $\{\phi_i\}_{i\geq 1}$ have been so chosen.

Let $f \in L^2(\tilde{D})$. Then f admits an eigenfunction expansion

$$f(x) = \sum_1^\infty a_i \phi_i(x), \qquad \text{a.e. } dx.$$

By Theorem 5.4 we have

$$e^{tH} f(x) = \sum_1^\infty e^{\lambda_i t} \phi[A_i(x), \qquad \text{a.e. } dx. \tag{5.7}$$

By Theorem 5.2 we can prove the following result.

5.6 PROPOSITION. *Let $f = \sum a_i \phi_i \in L^2(\tilde{D})$ and $t > 0$. Then*

$$\lim_{n\uparrow\infty} \sup_{x\in\tilde{D}} \sum_{i=n}^\infty |a_i e^{\lambda_i t} \phi_i(x)| = 0.$$

Consequently (5.7) can be improved as

$$e^{tH} f(x) = \sum_1^\infty a_i e^{\lambda_i t} \phi_i(x), \qquad \forall x \in \tilde{D}. \tag{5.8}$$

Moreover, from (5.8) we can prove that for any $(t,x,y) \in (0,\infty) \times \tilde{D} \times \tilde{D}$,

(5.9)
$$q(t,x,y) = \sum_1^\infty e^{\lambda_i t}\phi_i(x)\phi_i(y).$$

5.7 THEOREM.

(1) *Let $f \in L^2(\tilde{D})$ and λ be a real number. Then for any $x \in \tilde{D}$, the limit*

(5.10)
$$S_\lambda f(x) = \lim_{t\uparrow\infty} e^{-\lambda t} e^{tH} f(x)$$

always exists in $[-\infty,\infty]$. If $|S_\lambda f(x)| < \infty$ almost everywhere on $\tilde{D}$, then

(5.11)
$$\limsup_{t\uparrow\infty}\ \sup_{x\in\tilde{D}} e^{-\lambda t}|e^{tH} f(x)|$$

exists and is finite.

(2) *Let $f \in L^2(\tilde{D})$ with $\|f\|_{L^2} > 0$. Define*

$$\beta_1 = \inf\{\lambda \in R : \limsup_{t\uparrow\infty}\ \sup_{x\in\tilde{D}} e^{-\lambda t}|e^{tH} f(x)| < \infty\}.$$

Then $S_{\beta_1} f$ is an eigenfunction of $(\Delta/2+\mu)|_{\tilde{D}}$ corresponding to the eigenvalue β_1.

PROOF: (1) Let $f = \sum_1^\infty a_i\phi_i$. Without loss of generality we can assume that there exists at least one positive integer j such that $\sum_{\lambda_i=\lambda_j} a_i\phi_i(x)$ is not equal to zero, otherwise we shall have $e^{-\lambda t}e^{tH}f(x) = 0$ for all $t > 0$ and the limit (5.8) exists trivially. Setting

$$k = \min\{j : \sum_{\lambda_i=\lambda_j} a_i\phi_i(x) \neq 0\},$$

we have for $0 < \epsilon < t$,

(5.12)
$$e^{-\lambda t}e^{tH} f(x) = e^{-\lambda t}\left(\sum_1^\infty a_i e^{\lambda_i t}\phi_i(x)\right) \cdot$$
$$e^{(\lambda_k-\lambda)(t-\epsilon)-\lambda\epsilon}\left(\sum_{\lambda_i=\lambda_k} e^{\lambda_i\epsilon}a_i\phi_i(x) + \sum_{\lambda_i<\lambda_k} e^{(\lambda_i-\lambda_k)(t-\epsilon)}e^{\lambda_i\epsilon}a_i\phi_i(x)\right)$$

Let

$$\beta = \min\{\lambda_k - \lambda_i : \lambda_i < \lambda_k\}.$$

Then $\beta > 0$ and

$$\sum_{\lambda_i<\lambda_k} e^{(\lambda_i-\lambda_k)(t-\epsilon)}|e^{\lambda_i\epsilon}a_i\phi_i(x)| \le e^{-\beta(t-\epsilon} \sup_{y\in\tilde{D}} \sum_1^\infty |e^{\lambda_i\epsilon}a_i\phi_i(y)|,$$

consequently by Proposition 5.6,

$$\lim_{t\uparrow\infty} \sum_{\lambda_i<\lambda_k} e^{(\lambda_i-\lambda_k)(t-\epsilon)}|e^{\lambda_i\epsilon}a_i\phi_i(x)| = 0. \tag{5.13}$$

From (5.12) and (5.13) we can conclude that

(a) if $\lambda_k - \lambda > 0$, then

$$\lim_{t\uparrow\infty} e^{-\lambda t}e^{tH}f(x) = \begin{cases} +\infty, & \text{for } \sum_{\lambda_i=\lambda_k} a_i\phi_i(x) > 0 \\ -\infty, & \text{for } \sum_{\lambda_i=\lambda_k} a_i\phi_i(x) < 0; \end{cases}$$

(b) if $\lambda_k - \lambda = 0$,

$$\lim_{t\uparrow\infty} e^{-\lambda t}e^{tH}f(x) = \sum_{\lambda_i=\lambda_k} a_i\phi_i(x);$$

(c) if $\lambda_k - \lambda < 0$, then

$$\lim_{t\uparrow\infty} e^{-\lambda t}e^{tH}f(x) = 0.$$

Thus (5.10) is proved. The rest of the assertions can also be derived from (5.12) and (5.13).

5.8 THEOREM. *The eigenspace corresponding to the first eigenvalue of* $(\Delta/2+\mu)|_{\tilde{D}}$ *is one-dimensional.*

PROOF: See [**26**].

5.9 PROPOSITION. *The first eigenfunction* ϕ_1 *of* $(\Delta/2+\mu)|_{\tilde{D}}$ *can be chosen to be strictly positive.*

PROOF: Let ϕ_1 be the first eigenfunction of $(\Delta/2+\mu)|_{\tilde{D}}$. There exists at least one $y_0 \in \tilde{D}$ such that $|\phi_1(y_0)| \neq 0$. By (5.9) we have for $\epsilon > 0$,

$$q(\epsilon, x, y_0) = \sum_1^\infty e^{\lambda_i\epsilon}\phi_i(y_0)\phi_i(x), \qquad \forall x \in \tilde{D}.$$

Applying Theorem 5.7 we get

$$e^{\lambda_1\epsilon}\phi_1(y_0)\phi_1(x) = \lim_{t\uparrow\infty} e^{-\lambda_1 t}e^{tH}q(\epsilon,\cdot,y_0)(x), \qquad \forall x \in \tilde{D},$$

which is obviously nonnegative. Consequently ϕ_1 can be chosen to be nonnegative. By (5.4) and Theorem 5.4, ϕ_1 is strictly positive on $\tilde{D}$.

From now on we shall always assume that ϕ_1 has already been so chosen.

5.10 PROPOSITION. *Let $\phi \in L^2(\tilde{D})$ be strictly positive on $\tilde{D}$. Then the following assertions are equivalent:*

(1) *For some $x \in \tilde{D}$,*

$$\lim_{n\uparrow\infty} \int_{\tilde{D}} q(n,x,y)\phi(y)\,dy = 0.$$

(2) *All the eigenvalues of the operator $(\Delta/2+\mu)|_{\tilde{D}}$ are strictly negative.*

(3) *There exist positive constants c and β such that*

$$\sup_{x\in\tilde{D}} E_x\left(e^{A_t} I_{\{t<\tau\}}\right) \le ce^{-\beta t}.$$

PROOF: Suppose that $\phi \in L^2(\tilde{D})$ admits a Fourier expansion $\phi = \sum_1^\infty a_i\phi_i$, then we have

$$e^{tH}\phi(x) = \sum_{i=1}^{\infty} e^{\lambda_i t} a_i \phi_i(x), \qquad \forall x \in \tilde{D}.$$

Since $\phi_1 > 0$ is strictly positive, therefore if $\phi > 0$ then $a_1 = \int_{\tilde{D}} \phi_1(x)\phi(x)\,dx > 0$. Thus from the proof of theorem 5.7 we know that (1) and (2) must be equivalent.

(2)$\Longleftrightarrow$(3) is clear.

5.11 PROPOSITION. *Let $\gamma \in GK^d(\tilde{D})$ and C_t be the continuous additive functional of X_t associated with γ. Then*

(1) *For each compact subset $S \subset \tilde{D}$,*

$$\lim_{t\downarrow 0} \sup_{x\in A} |E_x\left(e^{A_\tau} g(X_\tau) I_{\{\tau\le t\}}\right)| = 0.$$

(2) $\lim_{t\downarrow 0} \sup_{x\in\tilde{D}} |E_x \int_0^{t\wedge\tau} e^{A_s}\,dC_s| = 0.$

PROOF: (1) Set $c = \sup_{x\in I_1} |g(x)|$, $u_t(x) = E_x\left(e^{|A|_t} I_{\{\tau\le t\}}\right)$. Then we have

$$|E_x\left(e^{A_\tau} g(X_\tau) I_{\{\tau\le t\}}\right)| \le c\cdot u_t(x).$$

Applying Theorem 5.3 we can prove that $u_t \in bC(\tilde{D})$ and

$$\lim_{t\downarrow 0} u_t(x) = 0.$$

But it is easy to see that $u_t(x)$ is decreasing as $t \downarrow 0$, consequently $u_t \downarrow 0$ uniformly on each compact subset A of $\tilde{D}$, which proves (1).

(2) Set $\overline{A}_t = |A|_t + |C|_t$. Then

$$|E_x \int_0^{t\wedge\tau} e^{A_s}\,dC_s| \le E_x \int_0^{t\wedge\tau} e^{\overline{A}_s}\,d\overline{A}_s.$$

Since

$$1+\int_0^{t\wedge\tau} e^{\overline{A}_s}\, d\overline{A}_s = e^{\overline{A}_{t\wedge\tau}} = 1+\int_0^{t\wedge\tau} e^{\overline{A}_{t\wedge\tau}-\overline{A}_s}\, d\overline{A}_s$$

we know that

$$\begin{aligned}\limsup_{t\downarrow 0}\sup_{x\in\tilde{D}} \left|E_x\int_0^{t\wedge\tau} e^{A_s}\, dC_s\right| &\le \limsup_{t\downarrow 0}\sup_{x\in\tilde{D}} E_x\int_0^{t\wedge\tau} e^{\overline{A}_s}\, d\overline{A}_s \\ &= \limsup_{t\downarrow 0}\sup_{x\in\tilde{D}} E_x\int_0^{t\wedge\tau} E_{X_s} e^{\overline{A}_{t\wedge\tau-s}}\, d\overline{A}_s \\ &= 0.\end{aligned}$$

6. Probabilistic Treatment of the Mixed Boundary Value Problem.

In this section we are going to give a probabilistic treatment of the mixed boundary value problem (4.1). But before doing this, we must give a criterion for the finiteness of the following function:

$$G(x) = E_x\left(e^{A_\tau} I_{\{\tau<\infty\}} + \frac{1}{2}\int_0^\tau e^{A_s}\, dL_s\right), \qquad x\in\tilde{D}.$$

The function defined above is called the gauge function for μ.

6.1 THEOREM (GAUGE THEOREM). *The following assertions are equivalent:*

(1) *For some $x\in\tilde{D}$, $G(x)<\infty$.*

(2) *G is bounded on $\tilde{D}$.*

(3) *All the eigenvalues of $(\Delta/2+\mu)|_{\tilde{D}}$ are strictly negative.*

(4) *There exist positive constants c and β such that*

$$\sup_{x\in\tilde{D}} E_x\left(e^{A_t} I_{\{t<\tau\}}\right) \le ce^{-\beta t}.$$

PROOF: (1)$\Longrightarrow$(3). Define

$$A\phi(x) = E_x\left(e^{A_\tau} I_{\{\tau\le 1\}} + \int_0^{1\wedge\tau} e^{A_s}\, dL_s\right).$$

It is easy to see that $\phi(x)>0$ for every $x\in\tilde{D}$, and that ϕ is bounded on $\tilde{D}$. By the Markov property we conclude that

$$G(x) = \phi(x) + \sum_0^\infty \int_{\tilde{D}} q(n,x,y)\phi(y)dy. \tag{6.1}$$

Assume that (1) holds; then it follows from (6.1) that

$$\lim_{n\uparrow\infty}\int_{\tilde{D}} q(n,x,y)\phi(x)\, dy = 0$$

consequently by Proposition 5.9, (3) is valid.

(3)$\Longrightarrow$(2) Suppose now that (3) holds. Let $\alpha = \lambda_1/2$. By the proof of Theorem 5.7 we know that

$$\lim_{t\uparrow\infty} \sup_{x\in\tilde{D}} e^{\alpha t}|e^{tH}\phi(x)| < \infty.$$

Therefore by (6.1)

$$\begin{aligned} \|G\|_\infty &\le \|\phi\|_\infty + \sum_{n=1}^{\infty} e^{-\alpha n} \sup_{x\in\tilde{D}} e^{\alpha n}\left| \int_{\tilde{D}} q(n,x,y)\phi(y)\,dy\right| \\ &= \|\phi\|_\infty + \sum_{n=1}^{\infty} e^{-\alpha n} \sup_{\tilde{D}} |e^{tH}\phi(x)| \\ &< \infty, \end{aligned}$$

which proves (2).

(2)$\Longrightarrow$(1) is trivial.

(3)$\Longleftrightarrow$(4) is proved in Proposition 5.10.

As a consequence of the gauge theorem we can obtain the following result.

6.2 THEOREM. *Let $\gamma \in GK^d(\tilde{D})$ be positive and C_t be the positive continuous additive functional of X_t associated with γ. If the gauge function of μ is finite, then*

$$\sup_{x\in\tilde{D}} E_x \int_0^\tau e^{A_t}\,dC_t < \infty.$$

PROOF: Obviously we have

$$\begin{aligned} E_x \int_0^1 e^{A_{t\wedge\tau}}\,dC_{t\wedge\tau} &\le E_x \int_0^1 e^{|A|_{t\wedge\tau}}\,dC_{t\wedge\tau} \\ &\le E_x\left(e^{|A|_1} C_1\right) \\ &\le E_x\left(e^{|A|_1 + C_1}\right). \end{aligned}$$

Since $|A|_t + C_t$ is the continuous additive functional of X_t associated with the measure $|\mu| + \gamma \in GK^d(\tilde{D})$, Khas'minskii's lemma implies that

$$\sup_{x\in\tilde{D}} E_x \int_0^1 e^{A_{t\wedge\tau}}\,dC_{t\wedge\tau} \le \sup_{x\in\tilde{D}} E_x\left(e^{|A|_1+C_1}\right) < \infty.$$

Thus it follows from Theorem 6.1 and the additive property that

$$\begin{aligned} E_x \int_0^\tau e^{A_t}\,dC_t &= \sum_{n=0}^{\infty} E_x \int_n^{n+1} e^{A_{t\wedge\tau}}\,dC_{t\wedge\tau} \\ &= \sum_{n=0}^{\infty} E_x\left(I_{\{n<\tau\}} e^{A_n} E_{X_n} \int_0^1 e^{A_{t\wedge\tau}}\,dC_{t\wedge\tau}\right) \\ &\le c \cdot \sup_{x\in\tilde{D}} E_x \int_0^1 e^{A_{t\wedge\tau}}\,dC_{t\wedge\tau} \cdot \sum_{n=0}^{\infty} e^{-\beta n} \\ &< \infty, \end{aligned}$$

where c and β are specified by Theorem 6.1 (4).

Now we are in a position to give a probabilistic treatment to the mixed boundary value problem (4.1).

6.3 THEOREM. *If $G \not\equiv \infty$, then*

$$u(x) = E_x\left(e^{A_\tau} g(X_\tau) I_{\{\tau<\infty\}} + \int_0^\tau e^{A_t}\, dB_t\right), \qquad x \in \tilde{D}, \tag{6.2}$$

is the unique solution to the problem (4.1).

PROOF: Let u be defined by (6.2). Since $G \not\equiv \infty$, Theorem 6.1 and Theorem 6.2 imply that u is bounded on $\tilde{D}$. By the Markov property and the additive property we can show that for all $t > 0$ and $x \in \tilde{D}$

$$u(x) = E_x\left(e^{A_t} u(X_t) I_{\{t<\tau\}} + e^{A_\tau} g(X_\tau) I_{\{t\geq\tau\}} + \int_0^{t\wedge\tau} e^{A_s}\, dB_s\right). \tag{6.3}$$

Set

$$u_t(x) = E_x\left(e^{A_t} u(X_t) I_{\{t<\tau\}}\right),$$
$$\epsilon_t(x) = E_x\left(e^{A_\tau} g(X_\tau) I_{\{t\geq\tau\}} + \int_0^{t\wedge\tau} e^{A_s}\, dB_s\right).$$

By Theorem 5.3 we have $u_t \in bC(\tilde{D})$ for each $t > 0$. On the other hand by Proposition 5.10 we know that $\epsilon_t \to 0$ uniformly on each compact subset A of $\tilde{D}$. Consequently $u = u_t + \epsilon_t$ is continuous on $\tilde{D}$. Using stochastic calculus we can conclude from (6.3) that u satisfies the following relation for all $t > 0$ and $x \in \tilde{D}$

$$u(x) = E_x\left(u(X_t) I_{\{t<\tau\}} + g(X_\tau) I_{\{t\geq\tau\}} + B_{t\wedge\tau} + \int_0^{t\wedge\tau} u(X_s)\, dA_s\right).$$

Put

$$\xi(x) = E_x\left(g(X_\tau) I_{\{\tau<\infty\}}\right).$$

Then it follows from Proposition 4.10 that $\xi \in \mathcal{H}_r(\tilde{D})$. It is easy to see that ξ satisfies the following relation for all $t > 0$ and $x \in \tilde{D}$:

$$\xi(x) = E_x\left(\xi(X_t) I_{\{t<\tau\}} + g(X_\tau) I_{\{t\geq\tau\}}\right).$$

Define a function w on $\tilde{D}$ by $w(x) = u(x) - \xi(x)$. Then $w \in bC(\tilde{D})$ and

$$w(x) = E_x\left(w(X_t) I_{\{t<\tau\}} + B_{t\wedge\tau} + \int_0^{t\wedge\tau} u(X_s)\, dA_s\right). \tag{6.4}$$

It follows from Lemma 5.1.4 of [**16**] and (6.4) that for any $h \in bC(\tilde{D})$,

$$\lim_{t\downarrow 0} \int_{\tilde{D}} h(x)\left(\frac{w(x)-P_t^0 w(x)}{t}\right) dx = \int_{\tilde{D}} h\,d\nu + \int_{\tilde{D}} hu\,d\mu.$$

Thus it follows from Lemma 1.3.4 of [**16**] that $w \in H_0^1(\tilde{D})$ and that

$$\frac{1}{2}\int_{\tilde{D}} \nabla w \cdot \nabla h\,dx = \int_{\tilde{D}} h\,d\nu + \int_{\tilde{D}} hu\,d\mu, \qquad h \in C_0^2(\tilde{D}),$$

which implies $\Delta w = 2\nu I_D + 2u\mu I_D$ in the distributional sense. By Proposition 4.2 we know that $\partial w/\partial n|_{I_2} = 2u\mu I_{I_2} + 2\nu I_{I_2}$. Moreover it is easy to see that $\lim_{t\uparrow\tau} \xi(X_t) = g(X_\tau)$ on $\{\tau < \infty\}$ and $\lim_{t\uparrow\tau} w(X_t) = 0$. Thus $u = w + \xi$ satisfies (1)—(4) of Definition 4.11.

Conversely suppose that u satisfies (1)—(4) of Definition 4.11 and that $u \in bC(\tilde{D})$. For each relatively open subset A of $\bar{G}$ such that $A \subset \overline{A} \subset \tilde{D}$, by Proposition 4.9 and Proposition 4.6 we have for $x \in A$,

$$u(x) = E_x\left(u(X_{t\wedge T_A}) + B_{t\wedge T_A} + \int_0^{t\wedge T_A} u(X_s)\,dA_s\right)$$

where $T_A = \inf\{t > 0 : X_t \notin A\}$. Letting $A \uparrow \tilde{D}$ we get the following relation by virtue of (3) of Definition 4.11,

$$u(x) = E_x\left(u(X_t)I_{\{t<\tau\}} + g(X_\tau)I_{\{t\geq\tau\}} + B_{t\wedge\tau} + \int_0^{t\wedge\tau} u(X_s)\,dA_s\right). \tag{6.5}$$

Using stochastic calculus we can get from (6.5) the following relation

$$u(x) = E_x\left(e^{A_t}u(X_t)I_{\{t<\tau\}} + e^{A_\tau}g(X_\tau)I_{\{t\geq\tau\}} + \int_0^{t\wedge\tau} e^{A_s}\,dB_s\right).$$

Since $G \not\equiv \infty$, letting $t \uparrow \infty$ in (6.5) and applying Theorem 6.2 we can get (6.2), which completes the proof.

The above theorem only solves the mixed boundary value problem (4.1) in the case that the gauge function G is finite, but in general, the gauge function for μ may fail to be finite. In the remaining part of this section we are going to construct a probabilistic solution to problem (4.1) without assuming that the gauge function for μ is finite.

It follows from Theorem 5.2 that we can choose a real number b such that the eigenvalues of the $(\Delta/2 + \mu - b)|_{\tilde{D}}$ are all strictly negative. Fix such a b and put

(6.6)

$$W(x) = E_x\left(\exp(A_\tau - b\tau)g(X_\tau)I_{\{\tau<\infty\}} + \int_0^\tau \exp(A_t - bt)\,dB_t\right), \qquad x \in \tilde{D}.$$

By Theorem 6.2 we know that $W \in bC(\tilde{D})$. Applying Theorem 5.7 we can pick out some nonnegative eigenvalues of $(\Delta/2+\mu)|_{\tilde{D}}$ by the following procedure. Set

$$b_1 = \inf\{\lambda \le b : \lim_{t\uparrow\infty} \sup_{x\in\tilde{D}} e^{-\lambda t}|e^{tH}W(x)| < \infty\},$$

$$W_1 = \lim_{t\uparrow\infty} e^{-b_1 t}e^{tH}W(x),$$

$$b_k = \inf\{\lambda \le b_{k-1} : \lim_{t\uparrow\infty} \sup_{x\in\tilde{D}} e^{-\lambda t}|e^{tH}(W - \sum_{i=1}^{k-1} W_i)(x)| < \infty\},$$

$$W_k(x) = \lim_{t\uparrow\infty} e^{-b_k t}e^{tH}(W - \sum_{i=1}^{k-1} W_i)(x),$$

until $b_k < 0$. Because there are at most finite number of nonnegative eigenvalues of $(\Delta/2+\mu)|_{\tilde{D}}$, the procedure described above is finite.

6.4 THEOREM. *Let W, $\{b_i\}$ and $\{W_i\}$ be defined as above.*

(1) *The mixed boundary value problem (4.1) is solvable iff $b_i \neq 0$ for all i.*

(2) *If $b_i \neq 0$ for all i, then for each $x \in \tilde{D}$, the limit*

$$u(x) = \lim_{t\uparrow\infty} E_x\left(e^{A_t}\left(W - \sum_{b_i>0}\frac{b}{b_i}W_i\right)(X_t)I_{\{t<\tau\}} + e^{A_\tau}g(X_\tau)I_{\{t\ge\tau\}} + \int_0^{t\wedge\tau} e^{A_s}\,dB_s\right), \tag{6.7}$$

exists and $u \in bC(\tilde{D})$. Furthermore, u is the unique solution to (4.1) which is orthogonal to the null space of the operator $(\Delta/2+\mu)|_{\tilde{D}}$. In particular, if 0 is not an eigenvalue of $(\Delta/2+\mu)|_{\tilde{D}}$, then u is the unique solution to (4.1).

PROOF: (1) Suppose that (4.1) is solvable. Let v be a solution of (4.1) which is orthogonal to the null space of $(\Delta/2+\mu)|_{\tilde{D}}$. Define

$$w(x) = E_x\int_0^\tau be^{A_t - bt}v(X_s)\,ds, \qquad x \in \tilde{D}. \tag{6.8}$$

By Theorem 6.3. w is the unique solution to the following problem

$$\begin{cases} \left(\frac{\Delta}{2} + \mu I_D\right) w + av = 0, & \text{in } D \\ w|_{I_1} = 0 \\ \frac{\partial w}{\partial n}|_{I_2} - 2w\mu I_{I_2} = 0 \end{cases}$$

Let $u = v - w$. By linearity, u is a solution to the following problem

$$\begin{cases} \left(\frac{\Delta}{2} + \mu I_D - b\right) u + \nu I_D = 0, & \text{in } D \\ u|_{I_1} = g \\ \frac{\partial u}{\partial n}|_{I_2} - 2u\mu I_{I_2} = 2\nu I_{I_2}. \end{cases}$$

Since the solution to the above problem is unique, we get $u = W$ where W is defined by (6.6). From (6.8) we know that w can be expressed as

$$w(x) = \lim_{t\uparrow\infty} \int_0^t be^{-bs} e^{sH} v(x)\, ds.$$

Therefore w is orthogonal to the null space of $(\Delta/2+\mu)|_{\tilde{D}}$ since so does v. To sum up the above, we conclude that W is orthogonal to the null space of $(\Delta/2+\mu)|_{\tilde{D}}$, which is equivalent to $b_i \neq 0$ for all i.

Conversely, suppose that $b_i \neq 0$ for all i, then in (2) we shall prove that the function u defined by (6.7) is a solution to (4.1). Hence (4.1) is solvable.

(2) Set

$$M_t = e^{A_t - bt} W(X_t) I\{t < \tau\} + e^{A_\tau - b\tau} g(X_\tau) I_{\{t \geq \tau\}} + \int_0^{t\wedge\tau} e^{A_s - bs}\, dB_s,$$
$$M_t^i = e^{-b_i t} e^{A_t} W_i(X_t) I_{\{t<\tau\}}.$$

Then it follows from Theorem 5.4 and the proof of Theorem 6.3 that $\{M_t^i\}_{t\geq 0}$, $\{M_t\}_{t\geq 0}$ are continuous P_x-martingales for each $x \in \tilde{D}$. Write

$$V(x) = W(x) - \sum_{b_i > 0} \frac{b}{b_i} W_i(x)$$

and

$$N_t = e^{A_t} V(X_t) I_{\{t<\tau\}} + e^{A_\tau} g(X_\tau) I_{\{t\geq\tau\}} + \int_0^{t\wedge\tau} e^{A_s}\, dB_s.$$

Applying stochastic calculus we get

$$N_t = N_0 + \bar{N}_t + \int_0^t e^{A_s} Z(X_s) I_{\{s<\tau\}}\, ds,$$

where $\bar{N}_t$ is a P_x-martingale null at 0 for each $x \in \tilde{D}$ and

$$Z(x) = b(W(x) - \sum_{b_i>0} W_i(x)).$$

For $0 \leq t < T < \infty$, we have

$$\int_t^T E_x |e^{A_s} Z(X_s) I_{\{s<\tau\}}|\, ds < \infty, \qquad \forall x \in \tilde{D},$$

thus by Fubini's theorem

$$E_x N_T - E_x N_t = E_x \int_t^T e^{A_s} Z(X_s) I_{\{s<\tau\}}\, ds = \int_t^T e^{sH} Z(x)\, ds. \tag{6.9}$$

By the procedure described before the statement of this theorem, there exists a real number $b_k < 0$ for which

$$\lim_{t\uparrow\infty} \sup_{x\in\tilde{D}} e^{-b_k t} |e^{tH} Z(x)| < \infty.$$

Take t_0 large enough so that

$$M = \sup_{t\geq t_0} \sup_{x\in\tilde{D}} e^{-b_k t} |e^{tH} Z(x)| < \infty.$$

Then for $t_0 \leq t < T$,

$$\sup_{x\in\tilde{D}} |E_x N_T - E_x N_t| \leq \sup_{x\in\tilde{D}} \int_t^T |e^{sH} Z(x)|\, ds \leq M \frac{e^{-|b_k|t}}{|b_k|}. \tag{6.10}$$

Consequently the right side of (6.7) converges uniformly in $x \in \tilde{D}$. Thus $u(x)$ is well defined for each $x \in \tilde{D}$. By (6.9), u can be expressed as

$$u(x) = E_x \left(e^{A_t} V(X_t) I_{\{t<\tau\}} + e^{A_\tau} g(X_\tau) I_{\{t\geq\tau\}} + \int_0^{t\wedge\tau} e^{A_s}\, dB_s \right) + \lim_{T\uparrow\infty} \int_t^T e^{sH} Z(x)\, ds. \tag{6.11}$$

In particular, taking $t = 0$, we get

$$u(x) = V(x) + \lim_{T\uparrow\infty} \int_0^T e^{sH} Z(x)\, ds. \tag{6.12}$$

By the constructions of W and W_i's and taking into account the uniform convergence of

$$\lim_{\epsilon\downarrow 0} \int_\epsilon^T e^{sH} Z(x)\, ds,$$

(6.12) shows that $u \in bC(\tilde{D})$. Again by (6.11) we know that u is orthogonal to the null space of $(\Delta/2 + \mu)|_{\tilde{D}}$ since W is.

Now we are going to show that u is a solution to (4.1). By virtue of (6.10), we can apply the dominated convergence theorem to obtain

$$\begin{aligned} &E_x\left(e^{A_t}I_{\{t<\tau\}}\lim_{T\uparrow\infty}\int_0^T E_{X_t}\left(e^{A_s}Z(X_s)I_{\{s<\tau\}}\right)ds\right)\\ &= \lim_{T\uparrow\infty} E_x\left(e^{A_t}I_{\{t<\tau\}}\int_0^T E_{X_t}e^{A_s}Z(X_s)I_{\{s<\tau\}}\,ds\right). \end{aligned} \tag{6.13}$$

It follows from the Markov property that

$$E_x\left(e^{A_t}I_{\{t<\tau\}}\int_0^T E_{X_t}e^{A_s}Z(X_s)I_{\{s<\tau\}}\,ds\right) = \int_t^T E_x e^{A_s}Z(X_s)I_{\{s<\tau\}}\,ds. \tag{6.14}$$

Applying (6.13) and (6.14) to (6.12) we arrive at

$$\begin{aligned} &E_x\left(e^{A_t}u(X_t)I_{\{t<\tau\}} + e^{A_\tau}g(X_\tau)I_{\{t\geq\tau\}} + \int_0^{t\wedge\tau} e^{A_s}\,dB_s\right)\\ &= E_x\left(e^{A_t}V(X_t)I_{\{t<\tau\}} + e^{A_\tau}g(X_\tau)I_{\{t\geq\tau\}}\right) + \lim_{T\uparrow\infty}\int_t^T e^{sH}Z(x)\,ds. \end{aligned}$$

Therefore by (6.11)

$$u(x) = E_x\left(e^{A_t}u(X_t)I_{\{t<\tau\}} + e^{A_\tau}g(X_\tau)I_{\{t\geq\tau\}} + \int_0^{t\wedge\tau} e^{A_s}\,dB_s\right)$$

from which we can show, by repeating the arguments in the proof of Theorem 6.3, that u is a solution to (4.1).

The uniqueness is clear.

7 The Semilinear Boundary Value Problem.

In this section we are going to apply the results obtained in Section 6 to prove the existence of solutions to the following semilinear boundary value problem:

$$\begin{cases} -\left(\frac{\Delta}{2} + \mu I_D\right)u + f_1(u) = \nu I_D & \text{in } D\\ u|_{I_1} = g\\ \frac{\partial u}{\partial n}|_{I_2} - 2u\mu I_{I_2} + 2f_2(u) = 2\nu I_{I_2}. \end{cases} \tag{7.1}$$

where f_1 and f_2 are continuously differentiable functions on R^1.

7.1 Definition. *A bounded continuous function u on $\tilde{D}$ is said to be a solution to (7.1) if*

(1) $u \in \mathcal{H}_r(\tilde{D}) \oplus H_0^1(\tilde{D})$;

(2) *u satisfies*

$$-\left(\frac{\Delta}{2}+\mu I_D\right)u+f_1(u)=\nu I_D \qquad \text{in } D$$

in the distributional sense;

(3) *u satisfies*

$$\frac{\partial u}{\partial n}|_{I_2}-2u\mu I_{I_2}+2f_2(u)=2\nu I_{I_2}$$

in the sense of Definition 4.1;

(4) *for any* $x\in\tilde{D}$,

$$\lim_{t\uparrow\tau}u(X_t)=g(X_\tau),P_x \text{ a.s. on } \{\tau<\infty\}.$$

The following lemma is essential for the main results following it.

7.2 LEMMA. *Let* $\bar{\mu}\in GK^d(\tilde{D})$ *and* $\overline{A}_t$ *be the continuous additive functional of* X_t *associated with* $\bar{\mu}$. *If both the gauge function for* μ *and the gauge function for* $\mu+\bar{\mu}$ *are finite, then the function defined below*

$$u(x)=E_x\left(e^{A_\tau}g(X_\tau)I_{\{\tau<\infty\}}+\int_0^\infty e^{A_t}\,dB_t\right), \qquad x\in\tilde{D}, \tag{7.2}$$

satisfies the following relation for any $x\in\tilde{D}$:

(7.3)

$$u(x)=E_x\left(e^{(A+\overline{A})_\tau}g(X_\tau)I_{\{\tau<\infty\}}+\int_0^\tau e^{(A+\overline{A})_t}\,dB_t+\int_0^\tau e^{(A+\overline{A})_t}u(X_t)\,d\overline{A}_t\right).$$

PROOF: It follows from Theorem 6.3 that the function u defined by (7.2) is a solution to the problem (4.1), consequently u is a solution to the following problem

$$\begin{cases} -\left(\frac{\Delta}{2}+\mu I_D+\bar{\mu}I_D\right)u=\nu I_D-u\bar{\mu}I_D & \text{in } D\\ u|_{I_1}=g \\ \frac{\partial u}{\partial n}|_{I_2}-2u\mu I_{I_2}-2u\bar{\mu}I_{I_2}=2\nu I_{I_2}-2u\bar{\mu}I_{I_2}. \end{cases} \tag{7.4}$$

Applying Theorem 6.3 to the problem (7.4) we can conclude that u satisfies (7.3).

7.3 THEOREM. *Let* f_1 *and* f_2 *satisfy the following conditions*

$$sf_i(s)\ge 0, \qquad i=1,2,\forall s\in R^1.$$

If the gauge function for μ *is finite then (7.1) admits a solution.*

PROOF: Using Theorem 6.3 and Lemma 7.2, repeating the argument in [**18**], we can show that (7.1) admits a solution. For details, see [**25**].

7.4 THEOREM. *Let f_1 and f_2 satisfy the following conditions*

(1) $f_i(0) = 0, \quad i = 1, 2;$

(2) $f_i(s) \geq 0, \quad i = 1, 2, \forall s \in (0, \infty).$

If the gauge function for μ is finite and if ν and g are nonnegative, then (7.1) admits a nonnegative solution.

PROOF: Similar to that of Theorem 7.3.

7.5 THEOREM. *Let f_1 and f_2 satisfy the following conditions*

(1) $f_i(0) = 0, \quad i = 1, 2;$

(2) $f_i(s) \leq 0, \quad i = 1, 2, \forall s \in (-\infty, 0).$

If the gauge for μ is finite and if ν and g are nonpositive, then (7.1) admits a nonpositive solution.

PROOF: Similar to that of Theorem 7.3.

REFERENCES

1. Adams, R. A., "Sobolev spaces," Academic Press, New York, 1975.
2. Airault, H., *Rèsolotion stochastique d'un problèm de Dirichlet-Neumann*, C. R. Acad Sci. Paris **280** (1975), 781–784.
3. Airault, H., *Problèm de Dirichlet-Neumann étalés et fonctionelles multiplicatives associeés*, in "Seminair sur les équations aux deriveés partielles III, collége de France," 1974–1975.
4. Airault, H., *Perturbations singuliéres et solutions stochastiques de problèm de D. Neumann-Spencer*, J. Math. Pures Appl. **54** (1976).
5. Aizenman, M.; Simon, B., *Brownian motion and Harnack's inequality for Schrödinger operators*, Comm. Pure Appl. Math. **25** (1982), 209–273.
6. Blanchard, Ph.; Ma, Zhiming, *Semigroup of Schrödinger operators with potentials given by Radon measures*, BiBoS No. 262 (1987).
7. Blanchard, Ph.; Ma, Zhiming, *Smooth measures and Schrödinger semigroups*, BiBoS No 295 (1987).
8. Blanchard, Ph.; Ma, Zhiming, *New results on the Schrödinger semigroups with potentials given by signed smooth measures*, BiBoS No. 308 (1988).
9. Blumenthal, R. M.; Getoor, R. K., "Markov processes and potential theory," Academic Press, New York, 1968.
10. Brosamler, G. A., *A probabilistic solution of the Neumann problem*, Math. Scand. **38** (1976), 137–147.
11. Chung, K. L., "Lectures from Markov processes to Brownian Motion," Springer-Verlag, New York, 1980.
12. Chung, K. L.; Hsu, Pei, *Gauge theory for the Neumann problem*, in "Seminar on Stochastic Processes," Birkhaüser, Boston, 1984.
13. Chung K. L.; Rao K. M., *Feynman-Kac functional and Schrödinger equation*, in "Seminar on Stochastic Procesess," Birkhaüser, Boston, 1981.
14. Chung K. L.; Rao, K. M., *General gauge theorem for multiplicative functionals*, Trans. Amer. Math. Soc. **306** (1988), 819–836.
15. Dellacherie, C.; Meyer, P. A., "Probabilités et potentiels: theorie des martingales," Herman, Paris, 1980.
16. Fukushima, M., "Dirichlet forms and Markov processes," North-Holland, Amsterdam, 1980.
17. Fukushima, M., *On two class of smooth measures for symmetric Markov process*, in "Stochastic analysis, Lecture Notes in Math., No 1322," Springer-Verlag, New York, 1988.
18. Glover, J.; McKenna, P. J., *Solving semilinear partial differential equations with probabilistic potential theory*, Trans. Amer. Math. Soc. **290** (1985), 665–681.

19. Hsu, Pei, *Reflecting Brownian motion, boundary local time and the Neumann problem*, Thesis, Stanford University.
20. Hsu, Pei, *Probabilistic approach to the Neumann problem*, Comm. Pure Appl. Math. **38** (1985), 445–472.
21. Ikeda, N.; Watanabe, S., "Stochastic differential equations and diffusions," North-Holland, Amsterdam, 1980.
22. Ma, Zhiming, *Probabilistic treatment of the Schrödinger equation with infinite gauge*, Scientia Sinica Ser. A **30** (1987), 685–695.
23. Ma, Zhiming, *On the probabilistic approach to the boundary value problems*, BiBoS No 288 (1987).
24. Ma, Zhiming; Song, Renming, *Probabilistic approach to the Neumann problem with infinite gauge*, Acta Math. Appl. Sinica **4** (1988), 30–40.
25. Ma, Zhiming; Song, Renming, *Probabilistic approach to the semilinear boundary value problem*, preprint (1988).
26. Ma, Zhiming; Song, Renming, *On the Schrödinger operator with mixed boundary conditions and singular potential*, forthcoming.
27. Ma, Zhiming; Zhao, Zhongxin, *Truncated gauge and Schrödinger operator with both sign eigenvalues*, in "Seminar on Stochastic Processes," Birkhaüser, Boston, 1987.
28. Oshima, Y, "Lecture on Dirichlet spaces," preprint, 1988.
29. Papanicolauo, V. G., *The probabilistic solution of the third boundary value problem for the Schrödinger equation and its path integral representation*, Thesis, Stanford University (1988).
30. Port, S. C.; Stone, C. J., "Brownian motion and classical potential theory," Academic Press, New York, 1978.
31. Sato, K.; Ueno, T., *Multidimensional diffusions and Markov processes on the boundary*, J. Math. Kyoto Univ. **4–3** (1965), 529–605.
32. Simon, B., *Schrödinger Semigroups*, Bull. Amer. Math. Soc. **7** (1982), 447–526.
33. Song, Renming, *Probabilistic approach to the third boundary value problem*, preprint (1987).
34. Song, Renming, *Solving semilinear boundary value problems with stochastic calculus*, preprint (1988).
35. Song, Renming, *Probabilistic approach to a class of semilinear partial differential equations*, Chinese J. of Appl. Prob. Stat. **5-1** (1989), 61-69.
36. Treves, F., "Basic linear partial differential equations," Academic Press, New York, 1975.
37. Yan, Jiaan, "Introduction to martingales and stochastic Integrals," Shanghai, 1981.

Zhiming Ma, Institute of Applied Mathematics, Academica Sinica, Beijing, China
Renming Song, Department of Mathematics, 201 Walker Hall, University of Florida, Gainesville, FL32611, USA

STOCHASTIC VARIATIONAL PRINCIPLE OF SCHRÖDINGER PROCESSES

by

MASAO NAGASAWA

Diffusion processes considered by Schrödinger (1931), which will be called Schrödinger processes, are prescribed by a pair of functions

$$(1)\qquad \begin{aligned} \phi(t,x) &= e^{\alpha(t,x) + \beta(t,x)}, \\ \hat{\phi}(t,x) &= e^{\alpha(t,x) - \beta(t,x)}. \end{aligned}$$

The product of them

$$(2)\qquad \phi(t,x)\hat{\phi}(t,x)$$

is the distribution density of a Schrödinger process X_t, and the log-derivative of ϕ gives a typical term of its drift

$$(3)\qquad \underline{\underline{a}}(t,x) + \frac{\nabla\phi}{\phi}(t,x),$$

where appears an additional drift term $\underline{\underline{a}}(t,x)$ which should be specified in applications. In the background of Schrödinger's attempt, as we see it immediately, there is time reversal and hence duality with respect to $\phi\hat{\phi}(t,x)$. Actually the drift of the time reversal of X_t is given by

$$\{-\underline{\underline{a}}(t,x) + \frac{\nabla\hat{\phi}}{\hat{\phi}}(t,x)\}^{\#}, \tag{4}$$

where # denotes the time reversal transformation (see (9) below). The drift term $\underline{\underline{a}}(t,x)$ in (3) stands for a vector potential of the electromagnetic field in applications and is assumed, for simplicity, to be bounded and to have Hölder continuous derivatives in x. In addition we put a gauge condition

$$\nabla\cdot\underline{\underline{a}}(t,x) = 0, \tag{5}$$

which is technically needed but of no harm in applications, since Schrödinger processes are invariant under the gauge transformation (cf. Wakolbinger-Stummer (preprint)). Since $\phi(t,x)$ may vanish, the drift $\nabla\phi/\phi$ diverges at the nodal set of $\phi(t,x)$. Therefore, we consider processes on $D = \{(t,x): \phi(t,x) \neq 0,\ t \in [a,b]\}$, and assume $\phi \in C^{1,2}(D)$.

Schrödinger processes can be constructed under some integrability conditions on $\phi(t,x)$ by a transformation in terms of multiplicative functionals (cf. Nagasawa(1987), for different ways of construction see Zheng-Meyer(1982,84), Carlen(1984) and Carmona(1985)). Let $\{\Omega, Q, (F_t), X_t, t\in[a,b]\}$ be a Schrödinger process, where Q stands for

$$Q = \int \phi\hat{\phi}(a,x)\,dx\,\tilde{Q}_{(a,x)},$$

with $\tilde{Q}_{(a,x)}$ defined after (3.39) in Nagasawa(1987). Then Q is absolutely continuous with respect to the

Wiener measure with an initial distribution $\phi\hat{\phi}(a,x)dx$, X_t does not hit the nodal set of $\phi(t,x)$, and it satisfies a stochastic differential equation

$$(6)\quad X_t = X_a + B_t + \int_a^t \{\underline{\underline{a}}(v,X_v) + \frac{\nabla\phi}{\phi}(v,X_v)\}dv, \quad t \in [a,b],$$

where B_t , $t \in [a,b]$, is a Brownian motion on $\{\Omega, Q, (F_t)\}$.

In this note we shall characterize Schrödinger processes in terms of stochastic variational principle (cf. pioneering contributions to the subject by Yasue(1981,86), and Zheng-Meyer(1982), Zambrini(1986)). We consider a class of pairs of F_t-(resp. $\hat{F}_t$-) semimartingales Y_t (resp. $\hat{Y}_t$), which are given by

$$(7)\quad Y_t = Y_a + B_t + \int_a^t \{\underline{\underline{a}}(v,Y_v) + \underline{\underline{b}}(v,\cdot)\}dv, \quad t \in [a,b],$$

and

$$(8)\quad \hat{Y}_t = \hat{Y}_a + \hat{B}_t + \int_a^t \{-\underline{\underline{a}}^{\#}(v,\hat{Y}_v) + \hat{\underline{\underline{b}}}(v,\cdot)\}dv, \quad t \in [a,b],$$

where $B_t, t \in [a,b]$ (resp. $\hat{B}_t, t \in [a,b]$) is a d-dim. F_t-(resp. $\hat{F}_t$-) Brownian motion, Y_a (resp. $\hat{Y}_a$) is $\phi\hat{\phi}(a,x)$-(resp. $\phi\hat{\phi}(b,x)$-) distributed, and $\underline{\underline{a}}^{\#}(t,x)$ is defined from $\underline{\underline{a}}(t,x)$ with the time reversal transformation

$$(9)\quad f^{\#}(t,x) = f(a+b-t,x), \quad t \in [a,b].$$

The most essential assumption is that Y_t and $\hat{Y}_t$ are time reversal of each other, i.e.

$$(10)\quad \hat{Y}_t = Y_{a+b-t}, \quad t \in [a,b], \quad Q\text{-a.e.}.$$

It is well known (cf. Nagasawa(1961,64)) that in Markovian cases, i.e. $\underline{\underline{b}}(v,\cdot) = \underline{\underline{b}}(v,Y_v)$ and $\hat{\underline{\underline{b}}}(v,\cdot) = \hat{\underline{\underline{b}}}(v,\hat{Y}_v)$ with $\underline{\underline{b}}(t,x)$ and $\hat{\underline{\underline{b}}}(t,x)$, they satisfy a duality relation

$$\hat{\underline{\underline{b}}}^{\#}(t,x) = -\underline{\underline{b}}(t,x) + \frac{\nabla\rho^{Y}(t,x)}{\rho^{Y}(t,x)}, \tag{11}$$

where $\rho^{Y}(t,x)$ denotes the distribution density of Y_t. For non-Markovian cases the duality relation becomes slightly complicated (cf. Föllmer(1986)).

In this note we consider ϕ and $\hat{\phi}$ such that at the initial time $t = a$ (resp. at the terminal time $t = b$)

$$\begin{aligned} &\int |\log\phi(a,x)|\phi\hat{\phi}(a,x)dx < \infty, \\ &\int |\log\hat{\phi}(a,x)|\phi\hat{\phi}(a,x)dx < \infty, \end{aligned} \tag{12}$$

(resp. with b in place of a), and a class of admissible processes Y_t such that

$$Q[\int_a^b \underline{\underline{b}}(v,\cdot)^2 dv] < \infty, \quad Q[\int_a^b \hat{\underline{\underline{b}}}(v,\cdot)^2 dv] < \infty, \tag{13}$$

and

$$Q[\int_a^b |c(v,Y_v)|dv] < \infty, \tag{14}$$

where $c(s,x)$ is the reference potential of ϕ defined by

$$c(s,x) = -\frac{L\phi}{\phi}(s,x) \tag{15}$$

with a parabolic differential operator

(16) $$L = \frac{\partial}{\partial s} + \frac{1}{2}\Delta + \underline{\underline{a}}(s,x)\cdot\nabla,$$

and the nodal set of ϕ (resp. $\hat{\phi}$) is not attained by the space-time process (t,Y_t) (resp. $(t,\hat{Y}_t)$).

REMARK. If $c(s,x)$ is given in advance instead of ϕ, then we assume the existence of a unique non-negative solution ϕ of

$$L\phi + c\phi = 0.$$

DEFINITION 1. An action functional $I(Y)$, $Y \in H$ is defined by

(17) $$I(Y) = Q[\int_a^b \{\frac{1}{2}\,\frac{\underline{\underline{b}}(v,\cdot)^2 + \hat{\underline{\underline{b}}}^{\#}(v,\cdot)^2}{2} - c(v,Y_v)\}dv + \frac{1}{2}\log\frac{\phi}{\hat{\phi}}(a,Y_a) - \frac{1}{2}\log\frac{\phi}{\hat{\phi}}(b,Y_b)].$$

REMARK. According to Föllmer(1986) there exists the forward derivative DY_t (resp. the backward derivative D^*Y_t) which was defined by Nelson(1966). In terms of them

(18) $$\underline{\underline{b}}(t,\cdot)^2 = (DY_t - \underline{\underline{a}}(t,Y_t))^2,$$
$$\hat{\underline{\underline{b}}}^{\#}(t,\cdot) = (D^*Y_t - \underline{\underline{a}}(t,Y_t))^2,$$

and hence the action functional $I(Y)$ as defined at (17) coincides with the one given by Yasue(1981,86). It is clear that

(19) $$\frac{1}{2}\log\phi/\hat{\phi} = \beta,$$

where β is the one in (1). If we define a wave function by

(20) $$\psi(s,x) = e^{\alpha(s,x) + i\beta(s,x)}$$

with α and β in (1), then it satisfies the Schrödinger equation

(21) $$i\frac{\partial\psi}{\partial s} + \frac{1}{2}\Delta\psi + i\underline{\underline{a}}(s,x)\cdot\nabla\psi - V(s,x)\psi = 0,$$

where the potential function $V(s,x)$ is given by

(22) $$V = \Delta\alpha + (\nabla\alpha)^2 + c \text{ , or with } \beta$$

$$= -2\frac{\partial\beta}{\partial s} - (\nabla\beta)^2 - 2\underline{\underline{a}}\cdot\nabla\beta - c.$$

Hence, the term $\frac{1}{2}\log\phi/\hat{\phi}(s,x) = \beta(s,x)$ in the definition of the action functional $I(Y)$ at (17) is the phase function of Schrödinger's wave function $\psi = e^{\alpha+i\beta}$.

If the Schrödinger equation (21) is given in advance, and ϕ and $\hat{\phi}$ are defined in terms of α and β of the solution $\psi = e^{\alpha+i\beta}$, then it is natural to replace $c(s,x)$ in (17) by the potential $V(s,x)$ in (21). Let us denote by $I^V(Y)$ the action functional with $V(s,x)$ in place of $c(s,x)$ at (17). Then, clearly

(23) $$I(Y) = I^V(Y) + Q\left[\int_a^b \{\Delta\alpha(v,Y_v) + (\nabla\alpha(v,Y_v))^2\}dv\right].$$

LEMMA. Let $c(s,x)$ be the reference potential of ϕ as defined at (15). Then

$$\hat{L}(\hat{\phi}^{\#})/\hat{\phi}^{\#} = -c^{\#}, \tag{24}$$

where

$$\hat{L} = \frac{\partial}{\partial s} + \frac{1}{2}\Delta - \underline{\underline{a}}^{\#}(s,x)\cdot\nabla. \tag{25}$$

Proof. It is clear that

$$\hat{L}(\hat{\phi}^{\#})/\hat{\phi}^{\#} = (\tilde{L}\hat{\phi}/\hat{\phi})^{\#}, \tag{26}$$

where $\tilde{L}$ is the formal adjoint of L

$$\tilde{L} = -\frac{\partial}{\partial s} + \frac{1}{2}\Delta - \underline{\underline{a}}(s,x)\cdot\nabla. \tag{27}$$

Here we have employed the gauge condition (5). However, with $\rho = \phi\hat{\phi}$, we have

$$\begin{aligned} \tilde{L}\hat{\phi}/\hat{\phi} &= \frac{\phi}{\rho}\tilde{L}\left(\frac{\rho}{\phi}\right) \\ &= \frac{1}{\phi}\left\{\frac{\partial\phi}{\partial s} + \frac{1}{2}\Delta\phi + \underline{\underline{a}}\cdot\nabla\phi\right\} \\ &\quad + \frac{1}{\phi}\left\{-\frac{\partial\rho}{\partial s} + \frac{1}{2}\Delta\rho - \nabla\cdot\left[\left(\underline{\underline{a}} + \frac{\nabla\phi}{\phi}\right)\rho\right]\right\}, \end{aligned} \tag{28}$$

where the gauge condition (5) is employed. Since $\rho = \phi\hat{\phi}$ is the distribution density of the Schrödinger processes X_t, the second term on the right-hand side of (28) vanishes, and the first one is equal to $-c = L\phi/\phi$ by the definition. Thus we have (24) because of (26).

THEOREM 1. (i) Let $I(Y)$ be the action functional as defined at (17). Then

$$I(Y) = \frac{1}{2}Q\Big[\int_a^b \Big\{\frac{1}{2}\Big(\underline{\underline{b}}(v,\cdot) - \frac{\nabla\phi}{\phi}(v,Y_v)\Big)^2 + \frac{1}{2}\Big(\hat{\underline{\underline{b}}}^{\#}(v,\cdot) - \frac{\nabla\hat{\phi}}{\hat{\phi}}(v,Y_v)\Big)^2\Big\}dv\Big]. \tag{29}$$

(ii) The Shrödinger process X_t is the extremal which attains the minimum of the action functional:

$$0 = I(X) = \min_{Y\in H} I(Y). \tag{30}$$

Proof. Applying Itô's formula to $\log\phi(t,x)$, we have

$$\begin{aligned} &\log\phi(b,Y_b) - \log\phi(a,Y_a) \\ &= \int_a^b \frac{\nabla\phi}{\phi}\cdot dB_v + \int_a^b \Big\{\frac{1}{\phi}L\phi(v,Y_v) + \frac{1}{2}\underline{\underline{b}}(v,\cdot)^2\Big\}dv \\ &\quad - \int_a^b \frac{1}{2}\Big(\underline{\underline{b}}(v,\cdot) - \frac{\nabla\phi}{\phi}(v,Y_v)\Big)^2 dv. \end{aligned} \tag{31}$$

Analogously

$$\begin{aligned} &\log\hat{\phi}^{\#}(b,\hat{Y}_b) - \log\hat{\phi}^{\#}(a,\hat{Y}_a) \\ &= \int_a^b \frac{\nabla\hat{\phi}^{\#}}{\hat{\phi}^{\#}}\cdot d\hat{B}_v + \int_a^b \Big\{\frac{1}{\hat{\phi}^{\#}}\hat{L}(\hat{\phi}^{\#})(v,\hat{Y}_v) + \frac{1}{2}\hat{\underline{\underline{b}}}(v,\cdot)^2\Big\}dv \\ &\quad - \int_a^b \frac{1}{2}\Big(\hat{\underline{\underline{b}}}(v,\cdot) - \frac{\nabla\hat{\phi}^{\#}}{\hat{\phi}^{\#}}(v,\hat{Y}_v)\Big)^2 dv. \end{aligned} \tag{32}$$

By Lemma the second term on the right-hand side of (32) is equal to

(33) $$\int_a^b (-c^{\#}(v,\hat{Y}_v) + \tfrac{1}{2}\hat{\underline{\underline{b}}}(v,\cdot)^2)dv = \int_a^b (-c(v,Y_v) + \tfrac{1}{2}\hat{\underline{\underline{b}}}^{\#}(v,\cdot)^2)dv,$$

and the third term is equal to

(34) $$-\int_a^b \frac{1}{2}(\hat{\underline{\underline{b}}}^{\#}(v,\cdot) - \frac{\nabla\hat{\phi}}{\hat{\phi}}(v,Y_v))^2 dv.$$

Further

(35) $$Q[\log\hat{\phi}^{\#}(b,\hat{Y}_b)] = Q[\log\hat{\phi}(a,Y_a)], \quad \text{and}$$
$$Q[\log\hat{\phi}^{\#}(a,\hat{Y}_a)] = Q[\log\hat{\phi}(b,Y_b)].$$

Combining (31)+(32) with (33),(34) and (35), we have

$$Q[\int_a^b \{\frac{1}{2}\,\frac{\underline{\underline{b}}^2 + (\hat{\underline{\underline{b}}}^{\#})^2}{2} - c\}dv + \frac{1}{2}\log\frac{\phi}{\hat{\phi}}(a,Y_a) - \frac{1}{2}\log\frac{\phi}{\hat{\phi}}(b,Y_b)]$$
$$= \frac{1}{2}Q[\int_a^b \{\frac{1}{2}(\underline{\underline{b}} - \frac{\nabla\phi}{\phi})^2 + \frac{1}{2}(\hat{\underline{\underline{b}}}^{\#} - \frac{\nabla\hat{\phi}}{\hat{\phi}})^2\, dv.$$

To be precise we consider the process Y_t (resp. $\hat{Y}_t$) up to $b\wedge T_n$ (resp. $b\wedge\hat{T}_n$), where

$$T_n = \inf\{v \in [a,b]:\ \phi(v,Y_v) \leqq \frac{1}{n},\ \text{or}\ \int_a^v (\frac{\nabla\phi}{\phi})^2(u,Y_u)du \geqq n\}$$

(resp. $\hat{T}_n$ with $\hat{\phi}^{\#}$), and use

$$\lim_{n\to\infty} Q[\log\phi(b\wedge T_n, Y_{b\wedge T_n})] = Q[\log\phi(b,Y_b)],$$

$$\lim_{n\to\infty} Q[\log\hat{\phi}^{\#}(b\wedge\hat{T}_n, \hat{Y}_{b\wedge\hat{T}_n})] = Q[\log\hat{\phi}^{\#}(b,\hat{Y}_b)].$$

The second assertion of the theorem is clear from the first.

For the process $\{\Omega, Q_{(s,x)}, (F_t), X_t, t \in [a,b]\}$ we can define an action functional by

$$(36) \quad I_{s,x}(Y) = Q_{(s,x)}\Big[\int_s^T \{\tfrac{1}{2}\underline{\underline{b}}(v,\cdot)^2 - c(v,Y_v)\}dv + \log\phi(s,Y_s) - \log\phi(T,Y_T)\Big],$$

which is familiar as a cost function in the theory of stochastic control, where T denotes a first hitting time to a subset.

THEOREM 2. (i) Let $I_{s,x}(Y)$ be defined by (36). Then

$$(37) \quad I_{s,x}(Y) = Q_{(s,x)}\Big[\int_s^T \tfrac{1}{2}\{\underline{\underline{b}}(v,\cdot) - \frac{\nabla\phi}{\phi}(v,Y_v)\}^2 dv\Big].$$

(ii) The process X_t with the drift $\underline{\underline{a}} + \nabla\phi/\phi$ attains the minimum of $I_{s,x}(Y)$

$$0 = I_{s,x}(X) = \min_{Y \in H} I_{s,x}(Y),$$

where H is a class of semimartingales Y_t with s in place of a in (7) satisfying

$$Q_{(s,x)}[|\log\phi(T,Y_T)|] < \infty,$$

and

$$Q_{(s,x)}\Big[\int_s^T \{\underline{\underline{b}}(v,\cdot)^2 + |c(v,Y_v)|\}dv\Big] < \infty.$$

Proof can be carried over in the same way.

REMARK. See Wakolbinger(preprint) for another $I(Y)$.

References

Carlen,E.A.(1984) Conservative diffusions, Comm.Math.Phys. 94, 293-315.

Carmona,R.(1985) Probabilistic construction of Nelson processes, Taniguchi Symp. PMMP Katata, 5-81.

Föllmer,H.(1986) Time reversal on Wiener space, Bibos-Symp. "Stochastic processes in Math.Phys." Springer Lect. Notes Math. 1158, 119-129.

Nagasawa,M.(1961) The adjoint process of a diffusion process with reflecting barrier, Kodai Math.Sem.Rep. 13, 235-248.

Nagasawa,M.(1964) Time reversions of Markov processes, Nagoya Math.Jour. 24, 177-204.

Nagasawa,M.(1987) Transformations of diffusion and Schrödinger processes, (preprint, to appear in Probab.Th. Rel.Fields)

Nelson,E.(1966) Derivation of Schrödinger equation from Newtonian Mechanics, Phys.Rev. 150, 1076-1085.

Schrödinger,E.(1931) Ueber die Umkehrung der Naturgesetze, Sitzungsberichte der Preussischen Akademie der Wissenschaften Physikalisch-Mathematishe Klasse, 144-153.

Wakolbinger,A.,Stummer,W.(preprint) On Schrödinger processes and stochastic Newton equations.

Wakolbinger,A.(preprint) A simplified variational principle of Schrödinger processes.

Yasue,K.(1981) Stochastic calculus of variations, J.Funct. Anal. 41, 327-340.

Yasue,K.(1986) The least action principle in quantum theory, Soryusiron Kenkyu (Japanese).

Zambrini,J.C.(1986) Variational processes and stochastic versions of mechanics, J.Math.Phys. 27, 2307-2330.

Zambrini,J.C.(1986) Stochastic mechanics according to Schrödinger, Phs.Rev. A. 33, 1532-1548.

Zheng,W.A.,Meyer,P.A.(1982) Quelques resultats de "Mechanique stochastique", Sém.Prob. XVIII, Lect.Notes Math. 1059, Springer, 223-244.

Zheng,W,A.,Meyer,P.A.(1984) Sur la construction de certain diffusions, Sem.Prob. XX, Lect.Notes Math. 1204, Springer 334-337.

Masao Nagasawa

Institut für Angewandte Mathematik

Universität Zürich

Rämistrasse 74, CH-8001 Zürich

Switzerland

THE HIGH CONTACT PRINCIPLE IN OPTIMAL STOPPING AND STOCHASTIC WAVES

by

BERNT ØKSENDAL*

SUMMARY

The high contact principle in optimal stopping states that at the boundary ∂D of the continuation region D the reward function g has a smooth fit with the optimal expected reward function g^*, in the sense that

$$g = g^* \text{on } \partial D$$

$$\nabla g = \nabla g^* \text{on } \partial D$$

Thus this principle gives the crucial link between optimal stopping and free boundary problems.

If the system is described by Brownian - or geometric Brownian - motion then it is easy to prove the high contact principle. However, in the general case when the system is described by a diffusion in $\mathbb{R}^n$ all the proofs in the literature known to the author are long and complicated.

The purpose of this work is to show how a result by Dynkin and Vanderbei about stochastic waves can be used to give a short proof of the high contact principle. Moreover, this proof works under weaker conditions than known before.

---

* Research supported in part by NAVF (Norway), ref. D.93.10.00

§1. Introduction

The (diffusion) optimal stopping problem can be formulated as follows:

Let $(X_t, \mathcal{M}_t, Q^x, \theta_t)$ be an Ito diffusion, i.e. the (strong) solution of an Ito stochastic differential equation in $\mathbb{R}^n$:

$$dX_t = b(X_t)dt + \sigma(X_t)dB_t \tag{1.1}$$

where $b : \mathbb{R}^n \to \mathbb{R}^n$ and $\sigma : \mathbb{R}^n \to \mathbb{R}^{n\times m}$ (where $\mathbb{R}^{n\times m}$ denotes the $n \times m$-matrices with real entries) are given Lipschitz functions with at most linear growth and $(B_t, \mathcal{F}_t, P^x, \theta_t)$ denotes m-dimensional Brownian motion. Let g be a given non-negative (or lower bounded) continuous function on $\mathbb{R}^n$. The optimal stopping problem is to find g^* and an $\mathcal{F}_t$-stopping time τ^* such that

$$g^*(x) := \sup_\tau E^x[g(X_t)] = E^x[g(X_{\tau*})], \tag{1.2}$$

where E^x denotes expectation w.r.t. the law Q^x of X_t starting at x, the sup being taken over all $\mathcal{F}_t$-stopping times τ. (If $\tau(w) = \infty$ then $g(X_\tau(w))$ is interpreted as 0). (The similar problem involving "inf" instead of "sup" in (1.2) can be transformed to (1.2) by changing sign on g, at least if g is bounded.)

The function g is often called the reward function. Thus we interpret $g(X_\tau)$ as the reward obtained by stopping X_t at time $t = \tau$. The problem is therefore to find a stopping time τ^* which maximizes the expected reward and to find this maximal reward g^*.

A fundamental result in optimal stopping is that τ^* can be realized as the first exit time τ_D for X_t from a certain region D (see for example [10, Theorem 10.9] for details):

THEOREM A (Optimal stopping theorem)

Let $\hat{g}$ denote the least superharmonic majorant of g. Then

$$g^* = \hat{g}.$$

Define

$$D = \{x; g(x) < \hat{g}(x)\} \quad \text{(the continuation region)}.$$

Assume that $\tau_D := \inf \{t > 0; X_t \notin D\}$ is finite a.s. Q^x and that the family

$$\{g(X_\tau)\}_{\tau \leq \tau_D} \text{ is uniformly } Q^x-\text{integrable.}$$

(For example, it suffices that g is bounded). Then

$$g^*(x) = E^x[g(X_{\tau_D})], \tag{1.3}$$

so $\tau^* = \tau_D$ is optimal.

Thus $g^* = \hat{g}$ coincides with g outside D. The high contact principle states that - under certain conditions - the contact between g and g^* on ∂D is smooth, in the sense that

$$\lim_{\substack{x \to y \\ x \in D}} \nabla g^*(x) = \nabla g(y) \text{ for } y \in \partial D \tag{1.4}$$

This result is crucial for the connection between optimal stopping problems and free boundary problems.

The first time this principle was formulated seems to be in a paper by Samuelson [11], who studied the optimal time for selling an asset, if the utility (reward) obtained by selling at the time t and when the price is z was given by

$$g(t,z) = e^{-\beta t}(z-1)^+,$$

the time-price process being

$$X_t = (t, Z_t),$$

where Z_t is a one-dimensional geometric Brownian motion, i.e.

$$Z_t = z \exp(\alpha t + \beta B_t),$$

for some constants α, β and a Brownian motion B_t.

A rigorous proof of the high contact principle in this case was given by McKean [7], in the Appendix to the same article.

Subsequently the high contact principle for time-space Brownian/geometric Brownian motion has been studied by several authors. See e.g. Bather [1], Merton [8], Van Moerbeke [13]. A proof for more general one-dimensional diffusions has been given by Shiryaev [12].

For general diffusions in $\mathbb{R}^n$ less is known. The most general results seem to be due to A Friedman [6], and A. Bensoussan and J. L. Lions [2]. They prove regularity results for solutions of variational inequalities. The high contact principle then follows from the equivalence between optimal stopping problems and certain variational inequalities. For example, we mention the following result:

THEOREM B ([6], Theorem 8.1):

Suppose the generator A of X_t, which is given by

$$Af(x) = \sum_{i=1}^{n} b_i(x)\frac{\partial f}{\partial x_i} + \sum_{i,j=1}^{n} a_{ij}(x)\frac{\partial^2 f}{\partial x_i \partial x_j}, \tag{1.5}$$

where $a = [a_{i,j}] = \frac{1}{2}\sigma\sigma^T$ (σ^T being the transposed of σ) satisfies:

(1.6) A is uniformly elliptic, i.e. there exists $\beta > 0$ such that $\xi^t a\xi \geq \beta|\xi|^2$ for all $\xi \in \mathbb{R}^n$

(1.7) a_{ij} and $\dfrac{\partial a_{ij}}{\partial x_t}$ are bounded functions on $\mathbb{R}^n$

$$b_i(x) = \sum_{j=1}^{n} \frac{\partial a_{ij}}{\partial x_j}(x) \tag{1.8}$$

Let $W^{k,p,\mu}$ denote the set of functions u on $\mathbb{R}^n$ whose first k weak derivatives exists and belong to L^p_{loc} and whose norm

$$\| u \|_{k,p,\mu} = \{\sum_{|\alpha|\leq k} \int |e^{-\mu|x|} D^\alpha u(x)|^p dx\}^{1/p} \tag{1.9}$$

is finite. Then if $f, g \geq 0$ satisfy

$$(1.10) \qquad g \in W^{2,p,\mu} \cap W^{2,2,\mu}$$

$$(1.11) \qquad f \in W^{0,p,\mu} \cap W^{0,2,\mu} \text{ for some } p > n,$$

then the solution $V(x)$ of the optimal stopping problem

$$(1.12) \qquad V(x) := \inf_{\tau} E^x \big[\int_0^{\tau} f(X_t)dt + g(X_\tau) \big]$$

is continuous, ∇V is continuous and

$$(1.13) \qquad \frac{\partial V}{\partial x_i} = \frac{\partial g}{\partial x_i} \text{ on } S,$$

where

$$S := \{x : \; V(x) = g(x)\},$$

(It is also assumed that

$$T := \inf\{t > 0; X_t \in S\} < \infty \text{ a.s. } Q^x$$

for all x).

Note that although the statement of the problem (1.12) appears to be more general than our original problem (1.2), it can be reduced to our case by considering the diffusion Y_t given by

$$(1.14) \qquad dY_t = \begin{bmatrix} dX_t \\ dZ_t \end{bmatrix} = \begin{bmatrix} b(X_t) \\ f(X_t) \end{bmatrix} dt + \begin{bmatrix} \sigma(X_t) & 0 \\ 0 & 1 \end{bmatrix} d\tilde{B}_t, Y_0 = (z, x)$$

where $\tilde{B}_t = (B_1(t), \dots, B_m(t), B_{m+1}(t))$ is $(m+1)$-dimensional Brownian motion, and the optimal stopping problem

$$(1.15) \qquad h^*(y) = h^*(x, z) = \sup_{\tau} E^{x,z}[h(Y_\tau)],$$

with $h(x, z) = z + g(x)$. We will return to this in the end of § 3.

It is natural to ask if there is a more direct approach to the high contact principle than via the formidable machinery of variational inequalities. The purpose of this paper is to point out that such a direct approach exists. Moreover, it gives the high contact conclusion with weaker assumptions than what appears to be known earlier.

In § 2 we show how a simple argument can be used to give a weak but general version of the high contact principle. Then in § 3 we apply results about stochastic waves due to Dynkin and Vanderbei [5] to obtain a strong version.

§2. A weak but general version

In this section we show that the high contact principle is basically a consequence of the strong Markov property, modulo some (non-tractable) differentiability conditions. First we introduce some notation: (As usual C^k denotes the family of functions whose derivatives of order up to k are continuous). Suppose X_t, g, D are as in Theorem A. Assume that ∂D locally (in a relatively open subset W of ∂D) is the graph of a function

$$x_j = \eta(y), \text{ where } y = \hat{x} = (x_1, \ldots, x_{j-1}, x_j, \ldots, x_n) \in V \subset \mathbb{R}^{n-1}, \tag{2.1}$$

where V is open, and that D near W is situated *"below"* W viewed along the x_j-axis. We introduce the following perturbations of D at W: Let V_0 be an open set with $\bar{V}_0 \subset V$, and for each $\theta \in (-1,1)$ let $\alpha_\theta : V \to \mathbb{R}$ satisfy

(i) $\alpha_\theta = 0$ outside V_0

(ii) $\alpha_0 = 0$

(iii) $(\theta, y) \to \alpha_\theta(y)$ is C^1 on $(-1,1) \times V$ and
$\alpha'_\theta(y) = \frac{d}{d\theta}\alpha_\theta(y) > 0$ on $(-1,1) \times V_0$.

Let D_θ denote the domain obtained by replacing ∂D in W by the graph of

$$x_j = \eta(y) + \alpha_\theta(y) \; ; \; y \in V.$$

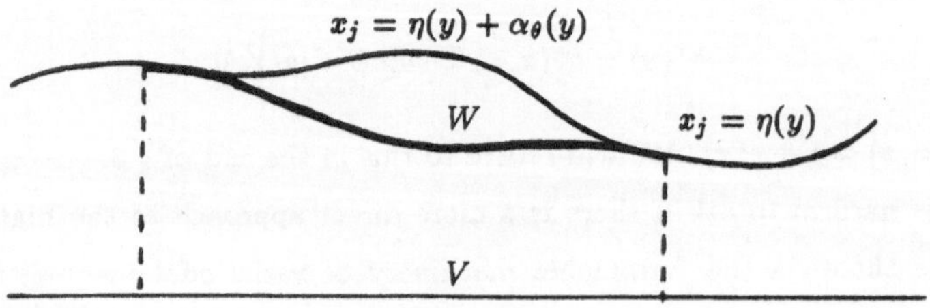

Let W_θ denote the corresponding part of ∂D_θ and let $\tau_\theta = \tau_{D_\theta}$ be the first exit time from D_θ (writing $\tau_0 = \tau$).

Finally, if h is a function on $\mathbb{R}^n$ we let $D_j^- h(x)$ denote the left-sided partial derivative of h with respect to x_j, i.e.

$$D_j^- h(x) = \lim_{\varepsilon \to 0+} \frac{g(x) - g(x - \varepsilon e_j)}{-\varepsilon}$$

(if the limit exists), where e_j is the jth unit vector in $\mathbb{R}^n$; $1 \le j \le n$.

THEOREM 1 (Weak version I of the high contact principle).

Let X_t, g, D be as in Theorem A. Suppose that, locally at $W \subset \partial D$, ∂D has the form (2.1) and D is situated below this graph there. Moreover, assume the following:

(2.2) The functions $x \to D_j g(x), x \to D_j g^*(x)$ are bounded and uniformly continuous in $G \cap D$, for some $\mathbb{R}^n$-neighborhood G of W

(2.3) All the points $x \in W$ are X-regular boundary points of D i.e. $P^x[\tau = 0] = 1$ for all $x \in W$.

(2.4) The function $F(\theta) = E^x[g(X_{\tau_\theta})]$ is differentiable at $\theta = 0$, for all $x \in D$

Then

$$\begin{aligned} &E^x[D_j^- g(X_\tau) \cdot \alpha_0'(\hat{X}_\tau) \cdot K_0] \\ &= E^x[D_j^- g^*(X_\tau) \cdot \alpha_0'(\hat{X}_\tau) \cdot K_0], x \in D, \end{aligned} \tag{2.5}$$

where

$$K_\theta = \chi_{X_{\tau_\theta} \in W_\theta} \text{ for } \theta \in (-1, 1),$$

χ being the indicator function (characteristic function).

<u>Proof.</u>

If $\theta < 0$ then $\tau_\theta < \tau$ so by the strong Markov property we have

$$
\begin{aligned}
(2.6)\qquad E^x[g(X_\tau)\cdot K_\theta] &= E^x[E^x[g(X_\tau)\cdot K_\theta|\mathcal{F}_{\tau_\theta}]\\
&= E^x\left[K_\theta\cdot E^{X_{\tau_\theta}}[g(X_\tau)]\right] = E^x[K_\theta\cdot g^*(X_{\tau_\theta})]
\end{aligned}
$$

This gives, with $\theta < 0$ and $\Delta\alpha = \alpha_0(\hat{X}_{\tau_\theta}) - \alpha_\theta(\hat{X}_{\tau_\theta})$,

$$
\begin{aligned}
F(0) - F(\theta) &= E^x[g(X_\tau)\cdot K_\theta] - E^x[g(X_{\tau_\theta})\cdot K_\theta]\\
&= E^x\left[(g(X_\tau) - g(X_{\tau_\theta} + \Delta\alpha\cdot e_j))K_\theta\right] + E^x\left[(g(X_{\tau_\theta} + \Delta\alpha\cdot e_j) - g(X_{\tau_\theta}))\cdot K_\theta\right]\\
&= -E^x[(g^*(X_{\tau_\theta} + \Delta\alpha\cdot e_j) - g^*(X_{\tau_\theta}))K_\theta] + E^x[(g(X_{\tau_\theta} + \Delta\alpha\cdot e_j) - g(X_{\tau_\theta}))\cdot K_\theta]\\
&= -E^x\left[D_j g^*(X_{\tau_\theta} + \mu\cdot e_j)\cdot\Delta\alpha\cdot K_\theta\right] + E^x[D_j g(X_{\tau_\theta} + \mu'\cdot e_j)\cdot\Delta\alpha\cdot K_\theta],
\end{aligned}
$$

where $0 \le \mu, \mu' \le \Delta\alpha$. Since $F(\theta)$ is maximal for $\theta = 0$ and F is differentiable at $\theta = 0$, we get from this that

$$
0 = \lim_{\theta\to 0^-}\frac{F(0) - F(\theta)}{-\theta} = -E^x[D_j^- g^*(X_\tau)\cdot\alpha_0'(\hat{X}_\tau)K_0] + E^x[D_j^- g(X_\tau)\cdot\alpha_0'(\hat{X}_\tau)\cdot K_0],
$$

as claimed.

In the previous result the condition (2.4) can be replaced by a one-sided differentiability condition on the X-harmonic extensions:

Define $\tilde{g}^\theta(x)$ to be the X-harmonic extension of $g|\partial D_\theta$ to D_θ, i.e.

$$
\tilde{g}^\theta(x) = E^x[g(X_{\tau_\theta})]\ ;\ \theta \in (-1, 1).
$$

THEOREM 2 (Weak version II of the high contact principle)

Let X_t, g, D be as in Theorem A. Suppose that, locally at $W \subset \partial D$, ∂D has the form (2.1) and D is situated below this graph there. Assume that there exists an $\mathbb{R}^n$-neighborhood G of W such that

$$
(2.7)\qquad x \to D_j g(x) \text{ is bounded and continuous in } G
$$

and

$$
(2.8)\qquad (\theta, x) \to D_j\tilde{g}^\theta(x) \text{ is uniformly continuous for } x \in G\cap D_\theta, \theta \in (-1, 1).
$$

Moreover, assume that all the points of W_θ are X-regular boundary points of D_θ, i.e.

$$
(2.9)\qquad P^x[\tau_\theta = 0] = 1 \text{ for all } x \in W_\theta, \theta \in (-1, 1).
$$

Then

$$E^x[D_jg(X_\tau)\cdot\alpha_0'(X_\tau)\cdot K_0]=E^x[D_j^-g^*(X_\tau)\cdot\alpha_0'(X_\tau)\cdot K_0], x\in D. \tag{2.10}$$

Proof. We proceed as in the proof of Theorem 1:
If $\theta_1<\theta_2$ then $\sigma:=\tau_{\theta_1}\le\zeta:=\tau_{\theta_2}$ so by the strong Markov property we get as in (2.6)

$$E^x[g(X_\zeta)\cdot K_{\theta_1}]=E^x[\tilde{g}^{\theta_2}(X_\sigma)\cdot K_{\theta_1}] \tag{2.11}$$

Therefore, if $\Delta\alpha=\alpha_{\theta_2}(\hat{X}_\sigma)-\alpha_{\theta_1}(\hat{X}_\sigma)$ the same argument as above gives

$$F(\theta_2)-F(\theta_1)=E^x[g(X_\zeta)]-E^x[g(X_\sigma)]$$

(2.12)

$$=-E^x[D_j\tilde{g}^{\theta_2}(X_\sigma+\xi e_j)\cdot\Delta\alpha\cdot K_{\theta_1}]+E^x[D_jg(X_\theta+\xi'\cdot e_j)\cdot\Delta\alpha\cdot K_{\theta_1}],$$

where $0\le\xi,\xi'\le\Delta\alpha$. In particular, F is a continuous function of θ.

Therefore, since $F(\theta)$ is maximal for $\theta=0$ we can find sequences

$$\theta_1(k)\le 0\le\theta_2(k)$$

such that

$$\theta_2(k)-\theta_1(k)\neq 0, \theta_2(k)-\theta_1(k)\to 0 \text{ and}$$

$$F(\theta_2(k))-F(\theta_1(k))=0 \text{ for all } k.$$

By (2.12) this gives, by dividing by $\theta_2(k)-\theta_1(k)$ and letting $k\to\infty$,

$$0=-E^x[D_j^-g^*(X_\tau)\cdot\alpha_0'(X_\tau)\cdot K_0]+E^x[D_jg(X_\tau)\cdot\alpha_0'(X_\tau)\cdot K_0],$$

which is (2.10).

Letting $x\to\partial D$ we deduce the usual conclusion of the high contact principle:

COROLLARY 1. Suppose the conditions of Theorem 1 or Theorem 2 are satisfied and that a point $\bar{x}\in W$ satisfies:

(2.13) There exists a sequence $\{x_k\}\subset D$ such that $x_k\to\bar{x}$ and $Q^{x_k}[|X_\tau-\bar{x}|<\varepsilon]\to 1$ as $k\to\infty$

for all $\varepsilon > 0$. Then

$$D_j g(\bar{x}) = \tilde{D_j} g^*(\bar{x})$$

<u>**REMARK**</u> Condition (2.13) holds for all $\bar{x} \in W$ if (for example) X_t is a strong Feller process. (See Dynkin [4, Th. 13.3]). And for this to be the case it is sufficient that the generator A of X_t is uniformly elliptic, i.e. that (1.6) holds.

§3. Application of stochastic waves

One of the major drawbacks of the approach in §2 was the intractable condition (2.4) (or (2.8)). We will now find conditions on the generator A which will ensure that (2.4) holds. This is achieved by applying a result of Dynkin and Vanderbei [5] about stochastic waves. We first summarize this theory. For details see [5].

Let φ be a real, measurable function on $\mathbb{R}^n$. For $t \geq 0$ define

$$T_t = inf\{s > 0; \quad \varphi(X_s) > \varphi(X_0) + t\}$$

Assume that

(3.1)
(i) $T_t < \infty$ for all t
(ii) $\varphi(X_t)$ is continuous in t
(iii) $T_0 = 0$ a.s. Q^x for all x.

Define $\tilde{X}_t = X_{T_t}, \tilde{\mathcal{M}}_t = \mathcal{M}_{T_t}$ and $\tilde{\theta}_t = \theta_{T_t}$. Then $(\tilde{X}_t, \tilde{\mathcal{M}}_t, Q^x, \tilde{\theta}_t)$ is a strong Markov process, called <u>the stochastic wave</u> corresponding to X and φ. The generator $\tilde{A}$ of $\tilde{X}_t$ is defined by

$$\tilde{A}f = \lim_{t \to 0} \frac{E^{\cdot}[f(\tilde{X}_t)] - f}{t}, \tag{3.2}$$

the limit being in the uniform topology on $\mathbb{R}^n$. We let $\tilde{\mathcal{D}}$ denote the set of functions f for which the limit (3.2) exists. We say that $f \in \tilde{\mathcal{D}}_x$ if there exists $f^* \in \tilde{\mathcal{D}}$ such that $f^* = f$ in a neighborhood of x in the topology generated by φ. In that case we put $\tilde{A}f(x) = \tilde{A}f^*(x)$.

If $V \subset \mathbb{R}^n$ is open let $C^{2,\lambda}(V)$ denote the set of functions with partial derivatives up to order 2 which are Hölder continuous with some exponent $\lambda > 0$. The main result of [5] then gives

THEOREM C [5]. Let φ satisfy (3.1) and assume that

(3.3) $$\varphi \in C^{2,\lambda}(\mathbb{R}^n)) \quad \text{and} \nabla \varphi \neq 0 \quad \text{in } \mathbb{R}^n,$$

(3.4) For some $x \in \mathbb{R}^n$ the set $G^x = \{y; \varphi(y) \leq \varphi(x)\}$ is bounded

(3.5) The generator A of X_t is uniformly elliptic (i.e. (1.6) holds).

Then every $f \in C^{2,\lambda}(\mathbb{R}^n)$ belongs to $\tilde{\mathcal{D}}_x$ and

(3.6) $$\tilde{A}f(x) = |\nabla \varphi(x)|^{-1}(\frac{\partial f}{\partial n}(x) + Hf(x)),$$

where

$Hf(x)$ is the interior normal derivative at x of the harmonic extension to G^x of $f|\partial G^x$ and $\frac{\partial f}{\partial n}$ denotes the derivative of f in the direction of the outer normal.

Localization

Before we apply this to the optimal stopping problem let us observe that we may localize the problem of high contact as follows:

Let G be a bounded open set and define $\hat{D} = G \cap D$. Modify g to a function g_1 satisfying the following conditions:

(a) $g_1 = g^*$ outside $\hat{D}$

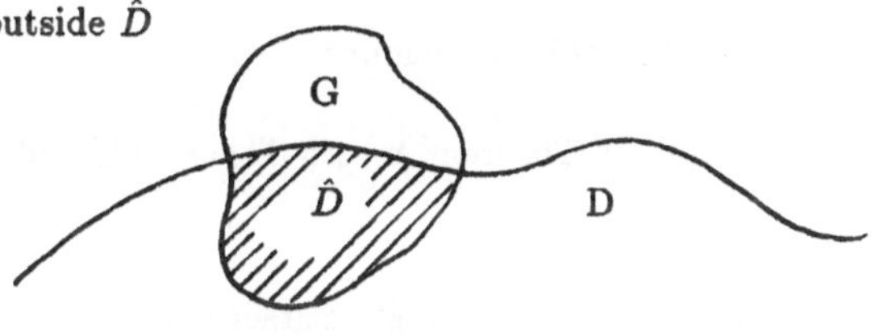

(b) $g_1 = g$ in a neighborhood of $\partial D \cap G$

(c) $g \leq g_1 < g^*$ on $\hat{D}$

(d) g_1 is continuous.

Then note that g^* is an X-superharmonic majorant of g_1 so if $\hat{g}_1$ denotes the least superharmonic majorant of g_1 ([10], Ch. X) we have

$$g_1^* \geq g^* \geq \hat{g}_1 = g_1^*$$

and we conclude that $g^* = g_1^*$. Consequently $\hat{D}$ is the continuation region for g_1. So if we want to investigate $D_j g$ and $D_j g^*$ near a point in ∂D we may reduce/modify D outside a neighborhood of this point as desired.

In particular, assume that, locally at $W \subset \partial D$, ∂D has the form (2.1), i.e.

$$x_j = \eta(\hat{x}) \quad ;\hat{x} = (X_1, \ldots, x_{j-1}, x_{j+1}, \ldots, x_n) \in V \subset I\!R^{n-1}$$

(with D situated "below" W) where we now assume that

$$\eta \in C^{2,\lambda}(V) \quad \text{for some} \lambda > 0 \tag{3.7}$$

Then by modifying D outside an $I\!R^n$-neighorhood of W we may assume that D has the form

$$D = \{x; \varphi(x) < 0\}, \tag{3.8}$$

where φ is a function in $C^{2,\lambda}(I\!R^n)$ such that

$$\varphi(x) = x_j = \eta(\hat{x}) \quad \text{for } x \in D \quad \text{near } W$$

and

$$\{y; \varphi(y) \le M\} \quad \text{is bounded for all } M, x$$

This enables us to combine Theorem A with Theorem C and we obtain the following:

THEOREM 3 (Strong version I of the high contact principle)

Let X_t, g, D be as in Theorem A. Assume that there exists $\lambda > 0$ such that, locally at $W \subset \partial D$, ∂D has the form

$$x_j = \eta(\hat{x}); \hat{x} = (x_1, \ldots, x_{j-1}, x_{j+1}, \ldots, x_n) \in V \subset I\!R^{n-1} \tag{3.9}$$

where $\eta \in C^{2,\lambda}(V)$ and let $\varphi \in C^{2,\lambda} I\!R^n)$ be the corresponding level function as in (3.8). Moreover, assume that

(3.10) The generator A of X_t is uniformly elliptic (i.e. (1.6) holds)

and

$$g \in C^{2,\lambda}(\mathbb{R}^n). \tag{3.11}$$

Then

$$D_j g(x) = \bar{D_j} g^*(x) \quad \text{for } x \in W.$$

<u>Proof</u>. By localization we may assume that D has the form

$$D = \{x; \varphi(x) < 0\}$$

(By uniform ellipticity $g^* \in C^{2,\lambda}(D)$ so the modified g_1 can be chosen in $C^{2,\lambda}(\mathbb{R}^n)$ too). By uniform ellipticity we obtain that (3.1) holds so we can apply Theorem C: We now consider the following analogue $H(t)$ of the function $F(\theta)$ of Theorem 1:

Define, for a fixed $x \in D$,

$$H(t) = E^x[g(X_{T_t})] = E^x[g(\tilde{X}_t)]; t \geq 0$$

By Dynkin's formula ([3, p. 133]) we have

$$H(t) = g(x) + E^x[\int_0^t \tilde{A}g(\tilde{X}_s)ds]$$

In particular, H is a differentiable function of t. Moreover, $H(t)$ is maximal for the value t_0 of t which gives $\varphi(x) + t_0 = 0$. Therefore $H'(t_0) = 0$.

Proceeding as in the proof of Theorem 1 we get for $t < t_0$

$$H(t_0) - H(t) = -E^x[\nabla g^*(X_{T_t} + \delta\gamma_t) \cdot \gamma_t] + E^x[\nabla g(X_{T_t} + \delta'\gamma_t) \cdot \gamma_t],$$

(for some $\delta, \delta' \in [0,1]$) where $\gamma_t = (t_0 - t) \cdot e_j$ near W, and elsewhere γ_t is the vector parallel to $\nabla\varphi(X_{T_t})$ such that

$$X_{T_t} + \gamma_t \in \partial D(\quad \text{i.e.} \gamma_t \approx (t_0 - t) \cdot \frac{\nabla\varphi}{|\nabla \varphi|^2})$$

Dividing by $t_0 - t$ and letting $t \uparrow t_0$ we obtain

$$\begin{aligned} 0 = E^x[(\frac{\partial g^*}{\partial n_i}(X_\tau) + \frac{\partial g}{\partial n_0}(X_\tau)) \cdot (1 - K_0)] \\ + E^x[(-\bar{D_j} g^*(X_\tau) + D_j g(X_\tau)) \cdot K_0], \end{aligned}$$

where $\frac{\partial}{\partial n_i}$ denotes the derivative in the direction of the inner normal to D. (It follows from Theorem 36.V in [9] that $\frac{\partial g^*}{\partial n_i}$ and $\bar{D_j} g^*$ exist and are bounded at ∂D).

Letting x in (3.12) approach a given point in W we obtain the conclusion of Theorem 3.

Finally we show how the more general optimal stopping problem

$$\gamma^*(x) = \sup_{\tau} E^x\left[\int_0^{\tau} f(X_s)ds + g(X_\tau)\right] \tag{3.13}$$

can be reduced to the case considered above. Let Y_t be as in (1.14), so that

$$Y_t = \begin{bmatrix} X_t \\ Z_t \end{bmatrix} \quad \text{with } Z_t = \int_0^t f(X_s)ds + B_{m+1}(t). \tag{3.14}$$

Define

$$h(y) = h(x, z) = z + g(x).$$

Then (since $E[B^0_{m+1}(\tau)] = 0$ for all τ)

$$\begin{aligned} h^*(y) &= \sup_{\tau} E^y[h(Y_\tau)] = \sup_{\tau} E^{x,z}[Z_\tau + g(X_\tau)] \\ &= \sup_{\tau} E^{x,z}[z + \int_0^{\tau} f(X_s)ds + B^0_{m+1}(\tau) + g(X_\tau)] \\ &= z + \sup_{\tau} E^x[\int_0^{\tau} f(X_s)ds + g(X_\tau)] = z + \gamma^*(x) \end{aligned}$$

If the generator A of X_t is uniformly elliptic, then so is the generator of Y_t. Therefore, if f is *"reasonable"* we can apply Theorem 3 to Y_t. We get the following conclusion:

THEOREM 4 (Strong version II of the high contact principle)

Let X_t and g be as in Theorem 3. Let f be a Lipschitz continuous function with at most linear growth. Define

$$\gamma^*(x) = \sup_{\tau} E^x[\int_0^{\tau} f(X_s)ds + g(X_\tau)],$$
$$D = \{x; g(x) < \gamma^*(x)\}.$$

Then at any point x of ∂D where ∂D locally has the form (3.9) of Theorem 3 we have

$$D_j g(x) = \bar{D_j} g^*(x)$$

Acknowledgement.

This paper was written while the author was visiting the University of California at San Diego (UCSD). I wish to thank UCSD and the Mathematics Department there for their hospitality.

REFERENCES

[1] J. A. Bather: Optimal stopping problems for brownian motion. Advances in Appl. Prob. **2** (1970), 259-286.

[2] A. Bensoussan & J. L. Lions: Applications of Variational Inequalities in Stochastic Control. North-Holland 1982.

[3] E. B. Dynkin: Markov Processes, Vol. I. Springer-Verlag 1965.

[4] E. B. Dynkin: Markov Processes, Vol. II. Springer-Verlag 1965.

[5] E. B. Dynkin & R. J. Vanderbei: Stochastic waves. Transactions Amer. Math. Soc. **275** (1983), 771-779.

[6] A. Friedman: Stochastic Differential Equations and Applications, Vol. II. Academic Press 1976.

[7] H. P. McKean: A free boundary problem for the heat equation arising from a problem of mathematical economics. Industrial managem. review **6** (1965), 32-39.

[8] R. C. Merton: The theory of rational option pricing. Bell J. of Economic & Management Science **4** (Spring) (1973), 141-183.

[9] C. Miranda: Partial Differential Equations of Elliptic Type. (2. ed.) Springer-Verlag 1970.

[10] B. Øksendal: Stochastic Differential Equations (2. ed.) Springer-Verlag 1989.

[11] P. A. Samuelson: Rational theory of warrant pricing. Industrial managem. review **6** (1965), 13-32.

[12] A. N. Shiryaev: Optimal Stopping Rules. Springer-Verlag 1978.

[13] P. Van Moerbeke: An optimal stopping problem with linear reward. Acta Mathematica **132** (1974), 111-151.

Current address:

Dept. of Mathematics
University of California, San Diego
La Jolla, CA 92093
USA

Dept. of Mathematics
University of Oslo
Box 1053 Blindern,
N-0316 Oslo 3
NORWAY

CONTINUITY OF SOLUTIONS OF SCHRÖDINGER EQUATION

by

Z.R. Pop-Stojanović and Murali Rao

__Introduction.__ In this paper we present a simpler proof of a Theorem due to M. Aizenman and B. Simon in [1] which states that the Schrödinger equation $(1/2)\Delta u + q\, u = 0$ in R^d, $d \geq 3$, $q \in K_d$, given in distributional sense, has a continuous solution in an open set D in R^d.

The question of giving a simpler proof of the above mentioned Theorem was recently posed by K. L. Chung.

__Setting.__ Let D be a nonempty open set in R^d, $d \geq 3$, and K_d be the Kato class of functions, i.e., $q \in K_d$, $d \geq 3$, if

$$\lim_{\alpha \downarrow 0} \sup_{x} \int_{|x-y| \leq \alpha} |x - y|^{-(d-2)} \, |q(y)| \, dy = 0 \quad .$$

Further notations and results used here are as follows. $(X_t)_{t\geq 0}$ will denote the standard Brownian motion in R^d. Let $\tau \equiv \tau_D = \inf\{s \geq 0;\ X_s \notin D\}$ denote the exit time of (X_t) from D. Recall that for any measurable f on ∂D, the function $E^x[f(X_\tau)]$ solves the Dirichlet problem for $(1/2)\Delta$, while the function

$$E^x[f(X_\tau)e^{-\int_0^\tau q(X_s)ds}]$$

solves the Dirichlet problem for $(1/2)\Delta + q$.

Now we can state the following theorem.

__Theorem.__ Suppose $u \in L^1_{loc}(D)$, $qu \in L^1_{loc}(D)$, and

(1) $$(1/2)\Delta u + q\,u = 0$$

in distribution sense in D. If $q \in K_d$, $d \geq 3$, then u is continuous in D.

<u>Proof.</u> The proof of this theorem is given in the following three steps.

<u>Step 1.</u> Taking a relatively compact open subset of D if necessary, we may assume $u \in L^1(D)$, $qu \in L^1(D)$. Let G be the Green function of D. Then Gqu makes sense (since $qu \in L^1(D)$), and by Young's inequality $Gqu \in L^p_{loc}$ for any $p < d/(d-2)$. Since $\Delta Gqu = -qu$, it follows that

(2) $$u = h + Gqu,$$

where h is a harmonic function in D. Since $Gqu \in L^p_{loc}(D)$ and h is harmonic and hence continuous, we see that $u \in L^p_{loc}(D)$, for any $p < d/(d-2)$. □

<u>Step 2.</u> The operator G_q defined by:

$$G_q f = Gqf = \int G(x,y)q(y)f(y)dy$$

maps $L^1(q)$ into itself. Indeed, if $\int |q(y)||f(y)|dy < \infty$, then

$$\| |q| Gqf \|_{L^1(D)} \leq \| G|q| \|_\infty \int |q(y)||f(y)|dy,$$

which follows from Fubini theorem. Since $u \in L^1(q)$ by assumption, and $Gqu \in L^1(q)$ from the above, it follows that $h \in L^1(q)$, where h is the harmonic function given in (2). □

<u>Step 3.</u> Now suppose $D = B(0,R)$ where $B(0,R)$ is a small ball in R^d such that $u \in L^p(D)$, $p < d/(d-2)$; $E^x[e^{p'\int_0^\tau q(X_s)ds}] < \infty$, and $\|G|q|\|_\infty < 1$. Since $\int_{B(0,R)} u^p = \int r^{d-1} \int_{\partial B(0,R)} |u(r,w)|^p dw < \infty$, one has

$$\int_{\partial B(0,R)} |u(r,w)|^p dw < \infty \quad \text{for almost all } r.$$

Choosing one such r, one gets $E^0[|u(X_\tau)|^p] < \infty$, and

(3) $$u = h + Gqu\ , \quad \|G|q|\|_\infty < 1, \quad h = E^x[u(X_\tau)].$$

Because of the second inequality in (3) and from Step 2, we see that the first equation in (3) has a unique solution given $h \in L^1(q)$. Now by the Feynman-Kac formula if $v = E^x[u(X_\tau)e^{-\int_0^\tau q(X_s)ds}]$, we know that v is locally bounded in D (by Hölder), and that v satisfies the equation

$$v = h + Gqv$$

where $h = E^x[u(X_\tau)]$. By the uniqueness of the solution it follows that v = u or that u is locally bounded. □

REFERENCES

[1] Aizenman, M. and Simon, B., Brownian motion and Harnack inequality for Schrödinger Operators, Commun. Pure and Applied Math., 35, pp. 209-273 (1982).

[2] Chung, K. L. and Rao, M., Feynman-Kac Functional and the Schrödinger Equation, Seminar on Stochastic Processes 1981, Birkhauser, Boston, pp. 1-29 (1981).

[3] Kato, T., Schrödinger Operators with Singular Potentials, Is. J. Math., 13, pp. 135-148 (1973).

DEPARTMENT OF MATHEMATICS, UNIVERSITY OF FLORIDA, GAINESVILLE, FL 32611.

STATIONARY SOLUTIONS FOR BILINEAR SYSTEMS WITH CONSTANT COEFFICIENTS

by

Gy. TERDIK

1. Introduction.

Recently an intensive study has been done for bilinear systems. There are many models in engineering, biology and socioeconomics, Mohler-Kolodziej(1980), which are nonlinear and/or nongaussian. Linear approximations do not describe them adequately. A large subclass of the nonlinear models, Gilbert(1978), Rugh(1981), which are possibly the closest to the linear ones, are the bilinear systems. These are defined as input-output systems which are linear with respect to either the input or the output, when one of them is fixed.

This paper deals with the continuous time, constant coefficient, bilinear systems driven by white noise input. A statistical study motivated by nuclear magnetic resonance spectroscopy and based on spectral analysis, was given by Brillinger(1989) for such systems. The results of this paper are pertinent to the continuation of the development of bilinear system identification procedures. We observe the weakly stationary (actually strictly stationary) bilinear process y_t as an L_2-functional in the Wiener space given by the input Wiener process v_t. The Wiener-Itô representation in terms of multiple Wiener-Itô integrals is given under the assumptions of physical realizability and stationarity, providing a natural generalization of the stochastic spectral representation in the Gaussian case. A necessary and sufficient condition for the stationarity follows from the structure of the transfer function system of the model. It is phrased in terms of the latent values of matrices of the system. Similar questions are investigated for discrete time bilinear systems in the papers Terdik(1985) and Terdik-Subba Rao(1989).

2. The Wiener-Itô Representation.

The bilinear system with single input v_t and single output y_t is given by the following state space equations

$$(2.1) \qquad d\mathbf{S}_t = (\mathbf{h} + A\mathbf{S}_t)dt + (\mathbf{b} + D\mathbf{S}_t)dv_t$$
$$y_t = \mathbf{c}'\mathbf{S}_t$$

where the input process v_t is standard Wiener process, the the state process $\mathbf{S}_t$ is vector valued and the A, D, $\mathbf{b}$, $\mathbf{c}$, $\mathbf{h}$ are constant appropriate matrices and vectors. Denote $\mathcal{B}_{\leq t}$ the σ- algebra generated by the process up to the time t.

We call the system (2.1) *physically realizable* if $\mathcal{B}_{\leq t}(\mathbf{S}) \subseteq \mathcal{B}_{\leq t}(v)$ i.e. the state process $\mathbf{S}_t$ and also the observation y_t depends only on the past of the noise.

The system (2.1) is called *subordinated* to the input v_t if $\mathbf{S}_t$ is measurable with respect to $\mathcal{B}(v)$ the σ-algebra generated by the input $v_t, t \in R$ having finite second moments and the time shift transformation for v_t is also time shift transformation for $\mathbf{S}_t$.

The following theorem is due to Dobrushin(1979) see also Major(1981).

Frequency Domain Representation Theorem. Let suppose that the process X_t is stationary in second order and subordinated to the Wiener process v_t then

$$(2.2) \qquad X_t = \sum_{0 \leq r} \frac{1}{r!} \int_{R^r} e^{it\Sigma\omega_{(r)}} g_r(\omega_{(r)}) \mathbf{W}(d\omega_{(r)})$$

where $\mathbf{W}(d\omega_{(1)})$ is the Wiener-Itô Random Measure (WIRM) corresponding to the $v_t, t \in R$, $E\mathbf{W}(d\omega_{(1)}) = 0$, $E|\mathbf{W}(d\omega_{(1)})|^2 = \frac{d\omega_1}{2\pi}$, the integrals are r-fold multiple Wiener-Itô integrals, the transfer functions $g_r(\omega_{(r)})$ are invariant under the permutations of its variables, $g_r(\omega_{(r)}) \in L_2^r$, $g_r(-\omega_{(r)}) = \overline{g_r(\omega_{(r)})}$ and $\Sigma\omega_{(r)} = \sum_{j=1}^r \omega_j$.

The representation (2.2) is unique and referred to as the *Wiener-Itô representation* for a stationary subordinated L_2 Wiener functional.
♣

Let us suppose now that there exists a *strictly* physically realizable i.e. subordinated and physically realizable solution for the state space equation (2.1). One can find sufficient conditions for this by Itô and Nisio(1964) in a very general circumstance. If the state process $\mathbf{S}_t$ is such a solution then it can be put into the form

$$(2.3) \qquad \mathbf{S}_t = \sum_{0 \leq r} \frac{1}{r!} \int_{R^r} e^{it\Sigma\omega_{(r)}} \mathbf{f}_r(\omega_{(r)}) \mathbf{W}(d\omega_{(r)})$$

where the transfer functions $\mathbf{f}_r(\omega_{(r)})$ are vector-valued and WIRM $\mathbf{W}(d\omega_{(1)})$ is connected to the input v_t by

$$v_t - v_s = \int_R \int_s^t e^{ih\omega_1} dh \mathbf{W}(d\omega_{(1)}). \tag{2.4}$$

We are going to put the representation (2.3) into the equation (2.1) to get an equation for the transfer functions. To do this we need the following

Lemma 1. Let $\mathbf{W}(d\omega_{(1)})$ be WIRM , $r \geq 1$ fixed, $g_r(\omega_{(r)}) \in L_2^r$, $g_r(-\omega_{(r)}) = \overline{g_r(\omega_{(r)})}$ and v_t defined by (2.4).

If

$$X_t = \int_{R^r} e^{it\Sigma\omega_{(r)}} g_r(\omega_{(r)}) \mathbf{W}(d\omega_{(r)}) \tag{2.5}$$

then

$$(i) \qquad \int_0^T X_t dt = \int_{R^r} g_r(\omega_{(r)}) \int_0^T e^{it\Sigma\omega_{(r)}} dt \mathbf{W}(d\omega_{(r)})$$

$$(ii) \qquad \int_0^T X_t dv_t = \int_{R^{r+1}} g_r(\omega_{(r)}) \int_0^T e^{it\Sigma\omega_{(r+1)}} dt \mathbf{W}(d\omega_{(r+1)}). \clubsuit$$

Proof. The formula (i) is an easy generalization of the well-known statement for stationary processes, see Gihman-Skorohod(1974).

To prove the equation (ii) one can use the Diagram Formula, see Major(1981), for simple processes and the increments (2.4) after that take the limit. The Diagram Formula is the following; let $f \in L_2^r$ and $h \in L_2$ then

$$\begin{aligned}\int_{R^r} f(\omega_{(r)}) \mathbf{W}(d\omega_{(r)}) \int_R h(\omega_1) \mathbf{W}(d\omega_{(1)}) \\ = \int_{R^{r+1}} f(\omega_{(r)}) h(\omega_{r+1}) \mathbf{W}(d\omega_{(r+1)}) \\ + \sum_{k=1}^{r} \int_{R^{r-1}} \int_R f(\omega_{(r)}) h(-\omega_k) \frac{d\omega_k}{2\pi} \mathbf{W}(d\omega_{(r)\setminus\{k\}}) \clubsuit\end{aligned}$$

where $\omega_{(r)\setminus\{k\}}$ denotes the vector $\omega_{(r)}$ when ω_k is missing. Now we are in the position to put the solution (2.3) into the equation (2.1) i.e. into

$$\int_0^T d\mathbf{S}_t = \int_0^T (\mathbf{h} + A\mathbf{S}_t)dt + \int_0^T (\mathbf{b} + D\mathbf{S}_t)dv_t$$

we get the the following equation for the rth order transfer function $\mathbf{f}_r(\omega_{(r)})$ by the Lemma 1, $r \geq 1$

$$\begin{aligned}
&\frac{1}{r!}\int_{R^r} (e^{iT\Sigma\omega_{(r)}} - 1)\mathbf{f}_r(\omega_{(r)})\mathbf{W}(d\omega_{(r)}) \\
&\quad = \frac{1}{r!}A\int_{R^r} \frac{(e^{iT\Sigma\omega_{(r)}} - 1)}{i\Sigma\omega_{(r)}}\mathbf{f}_r(\omega_{(r)})\mathbf{W}(d\omega_{(r)}) \\
(2.6) \qquad &\quad + \frac{1}{(r-1)!}D\int_{R^r} \frac{(e^{iT\Sigma\omega_{(r)}} - 1)}{i\Sigma\omega_{(r)}}\mathbf{f}_{r-1}(\omega_{(r-1)})\mathbf{W}(d\omega_{(r)}) \\
&\quad + \delta_{r=1}\mathbf{b}\int_{R} \frac{(e^{i\omega_1 T} - 1)}{i\omega_1}\mathbf{W}(d\omega_{(1)})
\end{aligned}$$

If we assume that all the eigenvalues of A are on the left halfplane which is the necessary and sufficient condition for the stationarity in linear case i.e. when D=0 then the equation (2.6) can be solved for $\mathbf{f}_r(\omega_{(r)})$ so that

$$\begin{aligned}
&\mathbf{f}_0 = -A^{-1}\mathbf{h} \\
(2.7) \qquad &\mathbf{f}_1(\omega_{(1)}) = (Ii\omega_1 - A)^{-1}(D\mathbf{f}_0 + \mathbf{b}) \\
&\mathbf{f}_r(\omega_{(r)}) = r(Ii\Sigma\omega_{(r)} - A)^{-1}D\mathbf{f}_{r-1}(\omega_{(r-1)}), \qquad r \geq 2
\end{aligned}$$

where I is the matrix of unity.

We have got the following

Theorem 1. If the bilinear system

$$d\mathbf{S}_t = (\mathbf{h} + A\mathbf{S}_t)dt + (\mathbf{b} + D\mathbf{S}_t)dv_t$$

$$y_t = \mathbf{c}'\mathbf{S}_t$$

has strictly physically realizable stationary solution and all the eigenvalues of A are on the left halfplane then the output process y_t can be given by the Wiener-Itô realization

$$y_t = \sum_{0 \leq r} \frac{1}{r!}\int_{R^r} e^{it\Sigma\omega_{(r)}} g_r(\omega_{(r)})\mathbf{W}(d\omega_{(r)})$$

where the transfer function system of the y_t is defined by

$$
\begin{aligned}
g_0 &= \mathbf{c}'A^{-1}\mathbf{h} \\
g_1(\omega_{(1)}) &= \mathbf{c}'(Ii\omega_1 - A)^{-1}(D\mathbf{f}_0 + \mathbf{b}) \qquad (2.8)\\
g_r(\omega_{(r)}) &= \mathbf{c}'(Ii\Sigma\omega_{(r)} - A)^{-1}D\mathbf{f}_{r-1}(\omega_{(r-1)}), \qquad r \geq 2. \clubsuit
\end{aligned}
$$

It is easy to see that there is a considerably jump between the scalar value bilinear state process and the vector valued one in the following sense. In the scalar valued case if the second order transfer function $\mathbf{f}_2(\omega_{(2)})$ is different from zero then for every $r \geq 2$, $\mathbf{f}_r(\omega_{(r)}) \neq 0$ i.e. the order of the transfer function system is either one (linear case) or infinite. The vector valued bilinear state process can produce a Nth $(N = 1, 2, \ldots)$ order transfer function system when $\mathbf{f}_N(\omega_{(N)}) \neq 0$ and $\mathbf{f}_r(\omega_{(r)}) = 0$ for every $r > N$. This process is called *degree-N polynomial model*, see Brillinger(1970). It is very likely that one of the simpliest nonlinear model is the second degree bilinear polynomial model.

Example 1. Put

$$
A = \begin{pmatrix} a_1 & 0 \\ 0 & a_2 \end{pmatrix}; \quad \mathbf{h} = \begin{pmatrix} 0 \\ 0 \end{pmatrix};
$$
$$
D = \begin{pmatrix} 0 & 0 \\ \delta & 0 \end{pmatrix}; \quad \mathbf{b} = \begin{pmatrix} 1 \\ 0 \end{pmatrix};
$$

where $a_1, a_2 < 0$, by the equation (2.1) the state space equations are

$$
\begin{aligned}
dS_t^{(1)} &= a_1 S_t^{(1)} dt + dv_t \\
dS_t^{(2)} &= a_2 S_t^{(2)} dt + \delta S_t^{(1)} dv_t
\end{aligned}
$$

The transfer functions are

$$
\mathbf{f}_1(\omega_{(1)}) = \begin{pmatrix} (i\omega_1 - a_1)^{-1} \\ 0 \end{pmatrix};
$$
$$
\mathbf{f}_2(\omega_{(2)}) = \delta \begin{pmatrix} 0 \\ (i\omega_1 - a_1)^{-1}(i\Sigma\omega_{(2)} - a_2)^{-1} \end{pmatrix}
$$

i.e. the state process $\mathbf{S}_t$ is a second order polynomial one.

3. Stationarity

The stationarity of finite order bilinear systems i.e. the degree-N polynomial ones assumes same necessary and sufficient assumptions as the linear systems. It is a consequence of the Lemma 2 that these systems have stationary solution iff all the eigenvalues of A are on the left halfplane.

From this point we assume that the bilinear system (2.1) is having infinite order. We have shown that the stationary solution of the state space equation (2.1) is

$$\mathbf{S}_t = \sum_{0\leq r} \int_{R^r} e^{it\Sigma\omega_{(r)}} \mathbf{f}_r(\omega_{(r)}) \mathbf{W}(d\omega_{(r)}) \tag{3.1}$$

where the transfer functions are given by

$$\begin{aligned} \mathbf{f}_0 &= -A^{-1}\mathbf{h} \\ \mathbf{f}_1(\omega_{(1)}) &= (Ii\omega_1 - A)^{-1}(D\mathbf{f}_0 + \mathbf{b}) \\ \mathbf{f}_r(\omega_{(r)}) &= (Ii\Sigma\omega_{(r)} - A)^{-1} D\mathbf{f}_{r-1}(\omega_{(r-1)}), \quad r \geq 2 \end{aligned} \tag{3.2}$$

It is easily follows from this that the necessary and sufficient conditions for the existence of the stationary solution for the bilinear state equation is that all the components of

$$\sum_{r=0}^{\infty} r! \int_{R^r} sym\mathbf{f}_r(\omega_{(r)}) \otimes \overline{sym\mathbf{f}_r(\omega_{(r)})} \mathbf{W}(d\omega_{(r)}) \tag{3.3}$$

be finite, where $\otimes$ stands for the usual tensor product of vector spaces. The $sym\mathbf{f}_r(\omega_{(r)})$ and also $\check{\mathbf{f}}_r(\omega_{(r)})$ denotes the symmetrized $\mathbf{f}_r(\omega_{(r)})$ i.e.

$$sym\mathbf{f}_r(\omega_{(r)}) = \frac{1}{r!} \sum_{P\in\mathcal{P}_r} \mathbf{f}(P\omega_{(r)}) = \check{\mathbf{f}}_r(\omega_{(r)}) \tag{3.4}$$

where $\mathcal{P}_r$ is the group of all permutations of the set $\{1, 2, \ldots, r\}$. The assumption on (2.3) gives assumptions for the parameters $\mathbf{h}$, A, $\mathbf{b}$ D of the system (2.1). Let us denote the $D \otimes D$ by $D^{\otimes 2}$.

Theorem 2. The bilinear system

$$d\mathbf{S}_t = (\mathbf{h} + A\mathbf{S}_t)dt + (\mathbf{b} + D\mathbf{S}_t)dv_t$$
$$y_t = \mathbf{c}'\mathbf{S}_t$$

has a stationary solution for every **c** if and only if all the eigenvalues of matrices A and

$$(I \otimes A + A \otimes I + D^{\otimes 2})^{-1}(D^{\otimes 2} - I \otimes A - A \otimes I)$$

have negative real parts and in this case the variance is

(3.5) $V^2 y_t = (\mathbf{c}' \otimes \mathbf{c}')[-I \otimes A - A \otimes I - D^{\otimes 2}]^{-1}(-DA^{-1}\mathbf{h} + \mathbf{b})^{\otimes 2}.$♣

Proof. The output y_t of the system is stationary for every **c** iff the state process $\mathbf{S}_t$ is stationary. To check the stationarity of the state process we need the following

Lemma 2. If $\mathbf{f}_r(\omega_{(r)})$ is defined by (3.2) and its symmetrized version is $\check{\mathbf{f}}_r(\omega_{(r)})$, see (3.4), and if all the eigenvalues of A are on the left halfplane then

$$H_r = r! \int_{-\infty}^{\infty} \cdots \int_{-\infty}^{\infty} \check{\mathbf{f}}_r(\omega_{(r)}) \otimes \overline{\check{\mathbf{f}}_r(\omega_{(r)})} \frac{d\omega_{(r)}}{(2\pi)^r}$$
$$= G^{r-1}\mathbf{f}^0; \quad r \geq 1$$

where

$$G = -(I \otimes A + A \otimes I)^{-1} D^{\otimes 2}$$
$$\mathbf{f}^0 = -(I \otimes A + A \otimes I)^{-1}(D\mathbf{f}_0 + \mathbf{b})^{\otimes 2}.$$♣

Proof. Let us begin with r=1. One gets by elementary integration that

$$H_1 = \int_{-\infty}^{\infty} (Ii\omega - A)^{-1} \otimes (-Ii\omega - A)^{-1} \frac{d\omega}{2\pi}(D\mathbf{f}_0 + \mathbf{b})^{\otimes 2}$$
$$= \mathbf{f}^0.$$

We note that for any appropriate matrices A, B, C, D the result

$$(AB) \otimes (CD) = (A \otimes C)(B \otimes D)$$

is valid.

The case r=2

$$H_2 = \frac{1}{2!}\int_{-\infty}^{\infty}\int_{-\infty}^{\infty} [(Ii\Sigma\omega_{(2)} - A)^{-1}D] \otimes [(-Ii\Sigma\omega_{(2)} - A)^{-1}D]$$

$$\times [(Ii\omega_1 - A)^{-1} \otimes (-Ii\omega_1 - A)^{-1}$$

$$+ (Ii\omega_1 - A)^{-1} \otimes (-Ii\omega_2 - A)^{-1}$$

$$+ (Ii\omega_2 - A)^{-1} \otimes (-Ii\omega_1 - A)^{-1}$$

$$+ (Ii\omega_2 - A)^{-1} \otimes (-Ii\omega_2 - A)^{-1}]\frac{d\omega_{(2)}}{(2\pi)^2}(D\mathbf{f}_0 + \mathbf{b})^{\otimes 2}$$

$$= \int_{-\infty}^{\infty}\int_{-\infty}^{\infty} [(Ii\Sigma\omega_{(2)} - A)^{-1}D] \otimes [(-Ii\Sigma\omega_{(2)} - A)^{-1}D]$$

$$+ (Ii\omega_2 - A)^{-1} \otimes (-Ii\omega_2 - A)^{-1}]\frac{d\omega_{(2)}}{(2\pi)^2}(D\mathbf{f}_0 + \mathbf{b})^{\otimes 2} = G\mathbf{f}^0$$

follows because if we integrate first by ω_2 for the mixed products we find that all the poles are in the upper half plane so by the Residue Theorem

$$\int_{-\infty}^{\infty}\int_{-\infty}^{\infty} [(Ii\Sigma\omega_{(2)} - A)^{-1}D] \otimes [(-Ii\Sigma\omega_{(2)} - A)^{-1}D]$$

$$\times [(Ii\omega_1 - A)^{-1} \otimes (-Ii\omega_2 - A)^{-1}]\frac{d\omega_{(2)}}{(2\pi)^2} = 0.$$

Consider now $r > 2$ and suppose that the Lemma 2 is valid up to r-1 by induction. Denote by $\omega_{(r)\setminus\{k\}}$ the vector $\omega_{(r)}$ when ω_k is missing and consider H_r

$$\frac{1}{r!}\int_{-\infty}^{\infty}\cdots\int_{-\infty}^{\infty} \check{\mathbf{f}}_r(\omega_{(r)}) \otimes \overline{\check{\mathbf{f}}_r(\omega_{(r)})}\frac{d\omega_{(r)}}{(2\pi)^r}$$

$$= \frac{1}{r!}\int_{-\infty}^{\infty}\cdots\int_{-\infty}^{\infty} [(Ii\Sigma\omega_{(r)} - A)^{-1}D] \otimes [(-Ii\Sigma\omega_{(r)} - A)^{-1}D]$$

$$\times (r-1)!^2 \sum_{k,l=1}^{r} \check{\mathbf{f}}_{r-1}(\omega_{(r)\setminus\{k\}}) \otimes \overline{\check{\mathbf{f}}_{r-1}(\omega_{(r)\setminus\{l\}})}\frac{d\omega_{(r)}}{(2\pi)^r}$$

$$= \frac{1}{r!}\sum_{k=1}^{r}\int_{-\infty}^{\infty}\cdots\int_{-\infty}^{\infty}[(Ii\Sigma\omega_{(r)} - A)^{-1}D]\otimes[(-Ii\Sigma\omega_{(r)} - A)^{-1}D]$$

$$\times (r-1)!^2 \check{\mathbf{f}}_{r-1}(\omega_{(r)\setminus\{k\}})\otimes\overline{\check{\mathbf{f}}_{r-1}(\omega_{(r)\setminus\{k\}})}\frac{d\omega_{(r)}}{(2\pi)^r}$$

$$+\frac{1}{r!}\int_{-\infty}^{\infty}\cdots\int_{-\infty}^{\infty}[(Ii\Sigma\omega_{(r)} - A)^{-1}D]\otimes[(-Ii\Sigma\omega_{(r)} - A)^{-1}D]$$

$$\times (r-1)!^2 \sum_{k\neq l}^{r}\check{\mathbf{f}}_{r-1}(\omega_{(r)\setminus\{k\}})\otimes\overline{\check{\mathbf{f}}_{r-1}(\omega_{(r)\setminus\{l\}})}\frac{d\omega_{(r)}}{(2\pi)^r}$$

$$= \frac{r}{r!}\int_{-\infty}^{\infty}\cdots\int_{-\infty}^{\infty}[(Ii\Sigma\omega_{(r)} - A)^{-1}D]\otimes[(-Ii\Sigma\omega_{(r)} - A)^{-1}D]$$

$$\times (r-1)!^2 \check{\mathbf{f}}_{r-1}(\omega_{(r-1)})\otimes\overline{\check{\mathbf{f}}_{r-1}(\omega_{(r-1)})}\frac{d\omega_{(r)}}{(2\pi)^r}$$

$$+\int_{-\infty}^{\infty}\cdots\int_{-\infty}^{\infty}[(Ii\Sigma\omega_{(r)} - A)^{-1}D]\otimes[(-Ii\Sigma\omega_{(r)} - A)^{-1}D]$$

$$\times \frac{(r-1)!^2}{(r-2)!}\check{\mathbf{f}}_{r-1}(\omega_{(r-1)})\otimes\overline{\check{\mathbf{f}}_{r-1}(\omega_{(r)\setminus\{r-1\}})}\frac{d\omega_{(r)}}{(2\pi)^r}$$

put $\omega_{(r-1)} = \omega_{(r-1)}$, $\mu = \Sigma\omega_{(r)}$ into the first term and integrate by μ. Then by induction we get the Lemma 2 except for showing that the second term is zero. Consider

$$\int_{-\infty}^{\infty}\cdots\int_{-\infty}^{\infty}[(Ii\Sigma\omega_{(r)} - A)^{-1}D]\otimes[(-Ii\Sigma\omega_{(r)} - A)^{-1}D]$$

$$\times \check{\mathbf{f}}_{r-1}(\omega_{(r-1)})\otimes\overline{\check{\mathbf{f}}_{r-1}(\omega_{(r)\setminus\{r-1\}})}\frac{d\omega_{(r)}}{(2\pi)^r}$$

$$= \int_{-\infty}^{\infty}\cdots\int_{-\infty}^{\infty}[(Ii\Sigma\omega_{(r)} - A)^{-1}D]\otimes[(-Ii\Sigma\omega_{(r)} - A)^{-1}D]$$

$$\times [(Ii\Sigma\omega_{(r-1)} - A)^{-1}D]\otimes[(-Ii\Sigma\omega_{(r)\setminus\{n-1\}} - A)^{-1}D]$$

$$\times \check{\mathbf{f}}_{r-2}(\omega_{(r-2)})\otimes\overline{\check{\mathbf{f}}_{r-2}(\omega_{(r)\setminus\{r-1,k\}})}\frac{d\omega_{(r)}}{(2\pi)^r}$$

put $\mu = \Sigma\omega_{(r)\setminus\{r-1\}}$ and integrate by μ. We get zero in the next step because all the poles of the integrand according to ω_{r-1} are in the upper half plane. ♣

To end the proof of Theorem 2 we have to remember only the fact that the eigenvalues of a matrix F have negative real parts iff the eigenvalues of the matrix $(I-F)^{-1}(I+F)$ are less then 1 in modulus. ♣

References

Brillinger, D.R.,(1970), The identification of polynomial systems by means of higher order spectra, Jour. Sound Vib. vol.12. pp. 301-314.

Brillinger, D.R.,(1989), A study of second- and third-order spectral procedures and maximum likelihood in the identification of a bilinear system, To appear.

Dobrushin, R.L.,(1979), Gaussian and their subordinated generalized fields, Ann. of Prob., 7, pp. 1-28.

Gihman, I.I., Skorohod, A.V.,(1974) The Theory of Stochastic Processes, Springer-Verlag, New-York.

Gilbert, E.,(1978), Bilinear and 2-power Input-output maps: Finite dimensional realizations and role of functional series, IEEE Trans. on Aut. Contr. vol. AC-23, pp. 418-425.

Itô, K., Nisio, M.,(1964), On stationary solutions of a stochastic differential equation, J. Math. Kyoto Univ., 4-1, pp. 1-75.

Mohler, R.R., Kolodziej, W.J., (1980), An overview of bilinear system theory and applications, IEEE Trans. on Syst., Man, and Cybern., vol. SMC-10, pp. 683-688.

Major, P., (1981), Multiple Wiener-Itô Integrals, Lecture Notes in Mathematics, Springer-Verlag, New York.

Rugh, W. J. (1981), Nonlinear System Theory, J. Hopkins Univ. Press, Baltimore.

Terdik, Gy., (1985), Transfer functions and conditions for stationarity of bilinear models with gaussian residuals, Proc. R. Soc. Lond. A 400, pp. 315-330.

Terdik, Gy., Subba Rao, T. (1987), On Wiener-Itô representation and the best linear predictors for bilinear time series, Techn. Rep. #187, Department of Mathematics, UMIST, Manchester, UK. To appear in Journ. of Appl. Prob.(1989, June).

György Terdik
Department of Statistics
UC Berkeley, CA 94720
Department of Mathematics
KLTE Pf. 12 Debrecen, Hungary, H 4010

Gaugeability for Unbounded Domains

by

Z. Zhao

Let D be a Greenian domain in $R^d (d \geq 1)$, namely, its Green function $G_D(x,y) < \infty$ for $x, y \in D$, $x \neq y$, and let $q \in K_d$ (see [1] for definition); if q is only given in D, then we assume $q(x) = 0$ for $x \in R^d - D$.

In a recent paper [1], Chung proved that if D is a domain with $m(D) < \infty$ (m is the Lebesgue measure) and $q \in K_d$, then $G_D|q|$ is bounded and converges to zero as $m(D) \to 0$. As Chung noticed later, the proof is valid if in the statements $m(D)$ is replaced by $M(D)$ defined below.

$$M(D) = \| G_D 1 \|_\infty = \sup_{x \in D} \int_D G_D(x,y) dy. \tag{1}$$

A domain with $M(D) < \infty$ is called Green-bounded. We re-state it as follows.

THEOREM 1. Let D be a Green-bounded domain in $R^d (d \geq 1)$ and $q \in K_d$. Then

(i) $G_D|q|$ is bounded in R^d;

(ii) if $M(D) \to 0$, then $G_D|q|$ converges to zero uniformly in R^d.

We have the inequality:

$$M(D) \leq A_d m(D)^{\frac{1}{d}}$$

where A_d is a constant depending on d only. Hence if $m(D) < \infty$ then D is Green-bounded.

Let us point out here that the proof also implies the following inequality. There exist constants C_0 and C_1 such that

$$\| G_D|q| \|_\infty \leq C_0 + C_1 \| G_D 1 \|_\infty .$$

Using Theorem 1 in the original version with $m(D)$ instead of $M(D)$, Chung proved the Gauge Theorem under the assumption that $m(D) < \infty$ and $q \in K_d$. He raised the question whether this can be generalized in the same way as Theorem 1, namely whether $M(D) < \infty$ and $q \in K_d$ is sufficient for the Gauge Theorem. We shall give a negative example here. On the other hand, his method can be extended without change to a special class of Green-bounded domains D:

$$\forall\, \epsilon > 0\ \exists \text{ a compact } K \subset D \text{ such that } M(D - K) < \epsilon. \quad (2)$$

i.e. the Gauge Theorem holds for a domain D satisfying (2) and $q \in K_d$. Clearly (2) is satisfied when $m(D) < \infty$.

We now present the negative example. We shall construct a Green-bounded domain D and a bounded function q such that the Gauge Theorem fails for (D, q), namely, the gauge u for (D, q) is finite and unbounded in D.

EXAMPLE Let $D = \{(x, y) \in R^2 : \ -\infty < x < \infty, -1 < y < 1\}$ and

$$q(x,y) = \frac{\left[\frac{\pi^2}{8}\ln(x^2+e) + \frac{x^2-e}{(x^2+e)^2}\right]\cos\left(\frac{\pi}{2}y\right)}{\ln(x^2+e)\cos\left(\frac{\pi}{2}y\right)+1}, (x,y) \in D.$$

(a) $M(D) < \infty$: For each $x, y \in D$,

$$E^{(x,y)}(\tau_D) = E^y\left(\tau_{(-1,1)}\right) = \int_{-1}^{1} G_{(-1,1)}(y,z)dz \leq 1.$$

(b) $0 \leq q \leq \dfrac{\pi^2}{8} + 1$:

Since $\ln(x^2+e) \geq 1$ and $\cos\left(\frac{\pi}{2}y\right) > 0,\ y \in (-1,1)$,

$$q(x,y) \geq \frac{\left(\frac{\pi^2}{8} - \frac{1}{e}\right)\cos\left(\frac{\pi}{2}y\right)}{\ln(x^2+e)\cos\left(\frac{\pi}{2}y\right)+1} > 0$$

and

$$q(x,y) \le \frac{\pi^2}{8} + \left|\frac{x^2 - e}{(x^2+e)^2}\right| \cos\left(\frac{\pi}{2}y\right) \le \frac{\pi^2}{8} + 1.$$

(c)

$$\begin{aligned} u((x,y)) &\equiv E^{(x,y)}\left\{\exp\left[\int_0^{\tau_D} q(X_t)dt\right]\right\} \\ &\le \ln(x^2+e)\cos\left(\frac{\pi}{2}y\right) + 1 < \infty, \text{ for } (x,y) \in D: \end{aligned}$$

Let

$$\varphi(x,y) = \ln(x^2+e)\cos\left(\frac{\pi}{2}y\right) + 1 \ge 1.$$

Then

$$\Delta\varphi(x,y) = \frac{2(e - x^2)}{(x^2+e)^2}\cos(\frac{\pi}{2}y) - \frac{\pi^2}{4}\ln(x^2+e)\cos(\frac{\pi}{2}y).$$

Hence we have

$$(\frac{\Delta}{2} + q)\varphi = 0.$$

i.e. φ is a positive solution in D.

For each $n \ge 1$, let

$$D_n = \left\{(x,y) \in R^2 : \quad -n < x < n, -1 + \frac{1}{n} < y < 1 - \frac{1}{n}\right\},$$

and clearly φ is a positive solution in $\overline{D}_n$. Hence by Theorem 1 in [3], we have

$$u_n((x,y)) \equiv E^{(x,y)}\left\{\exp\left[\int_0^{\tau_{D_n}} q(X_t)\,dt\right]\right\} \not\equiv \infty \quad \text{in } D_n.$$

Thus by Theorem 2 · 3 in [4], we have

$$\begin{aligned} \varphi(x,y) &= E^{(x,y)}\left\{\exp\left[\int_0^{\tau_{D_n}} q(X_t)\,dt\right]\varphi(X(\tau_{D_n}))\right\} \\ &\ge u_n((x,y)), \qquad \text{in } D_n. \end{aligned} \tag{3}$$

Since $\tau_{D_n} \uparrow \tau_D$ a.s., using Faton's lemma in (3) we obtain

$$u((x,y)) \le \varphi(x,y) < \infty, \quad (x,y) \in D.$$

(d) $u((x,y))$ is unbounded in D:

Since 0 is the first eigenvalue of $\left(\frac{1}{2}\frac{d^2}{dx^2}+\frac{\pi^2}{8}\right)$ in $(-1,1)$ with the Dirichlet boundary condition, we have by Theorem 1 in [3]:

$$E^0\left[e^{\frac{\pi^2}{8}\tau_{(-1,1)}}\right]=\infty.$$

For any given large number $N>0$, there exists a number $0<c<\frac{\pi^2}{8}$ such that

$$E^0\left[e^{c\tau_{(-1,1)}}\right]>N+2.$$

By the monotone convergence, we can find a number $0<\delta<1$ such that

$$E^0\left[e^{c\tau_{(-\delta,\delta)}}\right]>N+1.$$

Since $E^{(0,0)}\left[e^{c\,\tau_{(-\infty,\infty)\times(-\delta,\delta)}}\right]=E^0\left[e^{c\tau_{(-\delta,\delta)}}\right]$, there exists a number $A>0$ such that

$$E^{(0,0)}\left[e^{c\tau_{(-A,A)\times(-\delta,\delta)}}\right]>N.$$

It is easy to see that as $x\uparrow\infty, q(x,y)\to\frac{\pi^2}{8}$ uniformly for all $y\in(-\delta,\delta)$. Then there exists a number $L>0$ such that for all $x\geq L$ and $y\in(-\delta,\delta)$.

$$q(x,y)\geq c.$$

Since $q\geq 0$ in D, we have

$$\begin{aligned}E^{(L+A,0)}\left\{\exp\left[\int_0^{\tau_D}q(X_t)\,dt\right]\right\}&\geq E^{(L+A,0)}\left\{\exp\left[\int_0^{\tau_{(L,L+2A)\times(-\delta,\delta)}}q(X_t)dt\right]\right\}\\&\geq E^{(L+A,0)}\left\{\exp\left[c\tau_{(L,L+2A)\times(-\delta,\delta)}\right]\right\}\\&=E^{(0,0)}\left\{\exp\left[c\tau_{(-A,A)\times(-\delta,\delta)}\right]\right\}>N.\end{aligned}$$

Thus u is unbounded in D.

■

If D is unbounded and $q \in K_d$, then it follows from the Harnack's theorem (see [7]) that if we put

$$u(x) = E^x \left[\exp \int_0^{\tau_D} q(X_t)dt \right] \tag{4}$$

and assume $0 < u(x_0) < \infty$ for some x_0, then $0 < u(x) < \infty$ for all $x \in D$. Examples show that in $R^1, u(x)$ may not have a limit as $|x| \to \infty$, or it may tend to ∞, or a finite limit which may be 0. But $u(x)$ must converge to one as x approaches the only finite boundary point a in (a, ∞) or $(-\infty, a)$. This follows from Theorem 3 in Chung [2], but it can also be proved directly (Falkner gave a short proof). Recently Chung made the conjecture that the same is true in R^d for $d \geq 2$. Namely; Chung's Conjecture: Suppose D is unbounded, $q \in K_d$, and there exists $x_0 \in D$ such that $u(x_0) < \infty$. Then for any $z \in \partial D$, $\lim_{x \to z} u(x) = 1$ or at least u is bounded as $x \to z$, where ∂D does not contain ∞.

This conjecture is still open. However, in this paper it will be proved in the case where D is a Lipschitz domain. The condition is needed in order to apply the conditioned gauge theorem and the boundary Harnack principle. (see [5])

As usual, put

$$e_q(t) = \exp \left[\int_0^t q(X_s)ds \right], \qquad t \geq 0.$$

Theorem 2. Let D be a domain in R^d and $q \in K_d$ $(d \geq 1)$. Suppose $u(x) \equiv E^x [e_q(\tau_D)] < \infty$ in D and the boundary ∂D is locally Lipschitzian around a point $z \in \partial D$. Then we have

$$\lim_{x \to z} u(x) = 1 \tag{5}$$

Proof. For such a point $z \in \partial D$, there exists a number $r > 0$ sufficiently small such that $U = D \cap B(z, r)$ is a Lipschitz domain and $E^x [e_q(\tau_U)] < \infty$ in U, where $B(z,r)$ is the ball cencered at z with radius r. Then by the strong Markov property we have for each $x \in U$,

$$u(x) = E^x\left[\tau_U = \tau_D;\ e_q(\tau_D)\right] + E^x\left[\tau_U < \tau_D;\ e_q(\tau_U)u\left(X(\tau_U)\right)\right]. \tag{6}$$

Let $u_1(x)$ and $u_2(x)$ denote the first and the second term of the right side of (6), respectively. Then

$$u_1(x) = E^x\left[\tau_U = \tau_D;\ e_q(\tau_U)\right] = E^x\left[e_q(\tau_U)1_{\partial D \cap \partial U}(X(\tau_U))\right].$$

Since $1_{\partial D \cap \partial U}(y)$ on ∂U is continuous at z and z is a regular point, we have

$$\lim_{x \to z} u_1(x) = 1_{\partial D \cap \partial U}(z) = 1. \tag{7}$$

By the conditional gauge theorem for a bounded Lipschitz domain (see [5]), there exist $C_1 > 0$ and $C_2 > 0$ depending on U and q only, such that for all $x \in U$, $y \in \partial U$,

$$C_1 \le E_y^x\left[e_q\left(\tau_U\right)\right] \le C_2, \tag{8}$$

where E_y^x is the expectation on the conditioned Brownian motion starting with x and ending with y. (see [5]) Thus we have

$$u_2(x) = \int_{(\partial U) \cap D} E_y^x\left[e_q(\tau_U)\right] u(y) H_U(x, dy), \tag{9}$$

where $H_U(x, A)$ is the harmonic measure on ∂U, and then if we put

$$h(x) = \int_{D \cap \partial U} u(y) H_U(x, dy).$$

we get

$$C_1 h(x) \le u_2(x) \le C_2 h(x),\ x \in U. \tag{10}$$

Now take a sequence of bounded domains $\{D_n\}$ such that $D_n \uparrow\uparrow D$ and $D_1 \cap \partial U \neq \emptyset$. Put

$$h_n(x) = \int_{D_n \cap \partial U} u(y) H_U(x, dy), \; x \in U, \; n = 1, 2, \ldots .$$

For each $n \geq 1$, since $u(y)$ is bounded in $\overline{D}_n$, h_n is a strictly positive harmonic function in U and vanishes on $(\partial U) \cap (\partial D)$ continuously. Then by the boundary Harnack principle for a bounded Lipschitz domain (see e.g. [6]), there exists a number $C > 0$ depending on D, z and r only such that for any $x, x_0 \in D \cap B(z, \frac{r}{2})$ and $n \geq 1$,

$$\frac{h_n(x)}{h_1(x)} \leq C \frac{h_n(x_0)}{h_1(x_0)}. \tag{11}$$

Letting $n \to \infty$ we obtain by monotone convergence:

$$\frac{h(x)}{h_1(x)} \leq C \frac{h(x_0)}{h_1(x_0)}. \tag{12}$$

Thus by (10) and (12), we have

$$\begin{aligned} u_2(x) &\leq CC_2 \frac{h(x_0)}{h_1(x_0)} h_1(x) \leq \frac{CC_2}{C_1} \frac{u_2(x_0)}{h_1(x_0)} h_1(x) \\ &\leq \frac{CC_2}{C_1} \frac{u(x_0)}{h_1(x_0)} h_1(x). \end{aligned} \tag{13}$$

Now fix $x_0 \in D \cap B(z, \frac{r}{2})$ and let $x \to z$. Since $\lim_{x \to z} h_1(x) = 0$, we get by (13)

$$\lim_{x \to z} u_2(x) = 0. \tag{14}$$

Since $u = u_1 + u_2$, (5) follows from (7) and (14).

■

Corollary 4. Let D be a Lipschitz domain in R^d and $q \in K_d$ $(d \geq 1)$. If $u(x) \equiv E^x[e_q(\tau_D)] < \infty$ in D, then for any ball B in R^d, u is bounded in $D \cap B$.

The author would like to thank Professor K.L. Chung. He posed this problem and encouraged me to work on it.

References

[1] K.L. Chung, *Gauge Theorem for Unbounded Domains*, Seminar on Stochastic Processes 1988, 87–98.

[2] K.L. Chung, *On stopped Feynman-Kac functional*, Séminaire de Probabilités XIV (Univ. Strasbourg). Lecture Notes in Math. **784**, Springer-Verlag, Berlin, 1980.

[3] K.L. Chung, P. Li and R.J. Williams, *Comparison of probability and classical methods for the Schrödinger equation*, Exposition. Math. **4** (1986), 271–278.

[4] K.L. Chung and K.M. Rao, *Feynman-Kac functional and the Schrödinger equation*, Seminar on Stochastic Processes, 1981, 1–29.

[5] M. Cranston, E. Fabes and Z. Zhao, *Conditional gauge and potential theory for the Schrödinger operator*, Trans. Amer. Math. Soc. **307** (1988), 171–194.

[6] D. Jerison and C. Kenig, *Boundary behavior of harmonic functions in non-tangentially accessible domains*, Ann. of Math. **(2) 113** (1981), 367–382.

[7] Z. Zhao, *Conditional gauge with unbounded potential*, Z. Wahrsch. Vern. Gebiete, **65**, (1983) 13–18.

Z. Zhao
Department of Mathematics
University of Missouri
Columbia, MO 65211

CORRECTION TO: SOME FORMULAS FOR THE ENERGY FUNCTIONAL OF A MARKOV PROCESS

(Seminar on Stochastic Processes, 1988)

by

P. J. FITZSIMMONS and R. K. GETOOR

H. Kaspi has pointed out an error in the proof of Theorem 4.1: in line 3 on page 180 the inclusion $E_m \cap \{P_M u < \infty\} \subset \{u < \infty\}^r$ is *clearly* false. She has also provided a correct argument which we reproduce here. As noted in the first part of the paragraph preceding (4.19), $L(\xi, u) = \infty$ whenever $\xi(u = \infty) > 0$; similarly $L^M(\tilde{\xi}, \tilde{u}) = \infty$ whenever $\tilde{\xi}(\tilde{u} = \infty) > 0$. Thus to complete the proof of (4.2) it suffices to show that $\tilde{\xi}(\tilde{u} = \infty) > 0$ if $\xi(u = \infty) > 0$ and $L(\xi, P_M u) < \infty$. In this case $\xi(P_M u = \infty) = 0$, so

$$0 < \xi(u = \infty) = \xi(E_M \cap \{P_M u < \infty, u = \infty\}),$$

since $P_M u = u$ on $E \setminus E_M$. Also, $L(R_M \xi, u) = L(\xi, P_M u) < \infty$, and therefore $R_M \xi(u = \infty) = 0$. It follows that $\tilde{\xi}(E_M \cap \{P_M u < \infty, u = \infty\}) > 0$. But $\tilde{u} = u - P_M u$ on $E_M \cap \{P_M u < \infty\}$, hence $\tilde{\xi}(\tilde{u} = \infty) > 0$ as desired.

Department of Mathematics, C-012
University of California, San Diego
La Jolla, California 92093